Top 50 Skills for a Top Score: SAT Math

"What a surprise, what a relief! An SAT guide that actually meets you where you are, talks to you with wit and compassion, and clears away the panic of test-taking. And, the writing is first-rate too. Bravo Brian Leaf."

Rebecca Pepper Sinkler, former Editor, *The New York Times Book Review*

"I enjoyed the informal writing style, and the flash cards for math are brilliant! Students are used to stacks of vocabulary words in preparation for the verbal portion of the test, why not drills on flash cards for the math section?"

Denise Brown-Allen, Ed.D., Upper School Director, The Pingry School

"If everyone starts using Brian's secrets and strategies, The College Board and ETS are going to have to rewrite the SAT!!"

Max Shelton, George Washington University, Class of 2012

Top 50 Skills for a Top Score: SAT Critical Reading and Writing

"Brian Leaf has hacked off the head of America's high school boogie man—the dreaded SAT. He clearly lays out how the test works, accessible preparation strategies, and how to maximize one's score. Any college applicant can benefit from his thoughtful and well-researched advice."

Joie Jager-Hyman, former Assistant Director of Admissions, Dartmouth College, author of *Fat Envelope Frenzy: One Year, Five Promising Students and the Pursuit of the Ivy League Prize*

"A long time ago, in an era far, far away, I took the SAT—and I can remember the pressure and anxiety of it like it was yesterday. Lucky for you modern-day seniors, Brian Leaf has written the SAT guide to end all SAT guides. He thoroughly demystifies the test and lays out the 50 skills you need to max out your score. Better yet, Mr. Leaf writes with such humor, wit, and unpretentious expertise that you'll find yourself reading this book just for fun. I did. It almost—almost—even made me want to take the SAT again."

Sora Song, Senior Editor, *Time Magazine*

"What's more scary than facing SATs? Or more boring than prepping for them? For a student swinging wildly between angst and ennui, the solution is Brian Leaf's *Top 50 Skills for a Top Score: SAT Critical Reading and Writing*. Leaf, himself a genius at connecting with teenagers, meets students at their level, and spikes every drill with common sense and comedy. I especially loved the Superbad Vocabulary section—not your usual stuffy approach to language deficit disorder. Guaranteed to relax and engage the most reluctant (or panicked) student."

Rebecca Pepper Sinkler, former Editor, *The New York Times Book Review*

Top 50 Skills for a Top Score: ACT Math

"Anyone even thinking of taking the ACT needs this short but targeted guide to the math section. You simply can't afford not to spend the time reading his laser sharp drills that break down every type of problem on the ACT, show the math behind each type, and then provide drill sections based on that skill set. Even poor math students can learn to recognize all the types of math on the ACT and learn the ropes enough to get most of the easy and medium questions right every time. Mr. Leaf's guide is even entertaining as he gives the skill sets names like "Green Circle, Black Diamond" to make it feel like you are skiing rather than slogging through lessons. If you want a short but concise guide to the ACT with every trick and mathematical explanation necessary to get a perfect score, this is the book for you. You may even actually LEARN real math in the process as Mr. Leaf's love of the subject shines through so you don't just feel you are learning for a test."

Dr. Michele Hernandez, author of the bestselling books *A is for Admission, The Middle School Years,* **and** *Acing the College Application*

"Brian Leaf knows how to talk with students and in his book, *Top 50 Skills for a Top Score: ACT Math,* you can hear his voice loud and clear. Students who follow Brian's "Mantras" and work through the practice questions will gain confidence in their work, as well as improve their ACT scores."

Barbara Anastos, former Director, Monmouth Academy

"Feels like you have an insider divulging secrets from behind the walls of the ACT! At times going so far as to circumvent the math skills themselves, Brian gives practical tips and tricks specifically designed to outwit the ACT's formula, and he does it all with a sense of humor and fun. Nice job!"

Danica McKellar, actress (*The Wonder Years, West Wing***) and** *mathematician and author* **of** *New York Times* **bestsellers** *Math Doesn't Suck* **and** *Kiss My Math*

Top 50 Skills for a Top Score: ACT English, Reading, and Science

"This book is a good read even if you *don't* have to take the ACT."

Edward Fiske, author of the bestselling college guide, the *Fiske Guide to Colleges*

"The **specific** skills needed for the ACT, confidence building, stress-management, how to avoid careless errors . . . this book has it covered!"

Laura Frey, Director of College Counseling, Vermont Academy
Former President, New England Association for College Admission Counseling

McGraw-Hill Education
Top 50 Skills for a Top Score:
ACT Math

Second Edition

Brian Leaf, M.A.

New York | Chicago | San Francisco | Athens | London | Madrid

Mexico City | Milan | New Delhi | Singapore | Sydney | Toronto

2 3 4 5 6 7 8 9 0   CUS/CUS   1 2 1 0 9 8 7 6

ISBN   978-1-2595-8625-5
MHID   1-2595-8625-1

33614057706144

e-ISBN   978-1-2595-8626-2
e-MHID   1-2595-8626-X

ACT is a registered trademark of ACT, which was not involved in the production of, and does not endorse, this product.

McGraw-Hill Education books are available at special quantity discounts to use as premiums and sales promotions or for use in corporate training programs. To contact a representative, please visit the Contact Us pages at www.mhprofessional.com.

Contents

How to Use This Book

It's simple. The questions that will appear on your ACT are predictable. We know the recipe: a multiples question, an average question, a midpoint or distance question . . . And while each of these topics is broad and could be the subject of a whole mathematics course, **the ACT always tests the same concepts!**

In this book, I will teach you exactly what you need to know. I will introduce each topic and follow it with drills. After each set of drills, check your answers. Read and reread the solutions until they make sense. They are designed to simulate one-on-one tutoring, like I'm sitting right there with you. Reread the solutions until you could teach them to a friend. In fact, do that! My students call it "learning to channel their inner Brian Leaf." There is no better way to learn and master a concept than to teach it!

Any new concept that you master will be worth points toward your ACT math score. That's the plan; it is that simple. If you did not understand functions, ratios, and averages before, and you do now, you will earn 3+ extra points.

This book is filled with ACT Math Mantras. They tell you what to do and when to do it. "When you see a proportion, cross-multiply." "When you see a linear pair, determine the measures of the angles." This is the stuff that girl who got a 36 does automatically. The Mantras teach you to think like her.

"Sounds good, but the ACT is tricky," you say. It is, but we know their tricks. Imagine a football team that has great plays, but only a few of them. We could watch films and study those plays. No matter how tricky they are, we could learn them, expect them, and beat them. ACT prep works the same way. You will learn the strategies, expect the ACT's tricks, and raise your score. Now, go learn and rack up the points!

Easy, Medium, Hard, and Guessing

The ACT is not graded like a math test at school. If you got only half the questions right on an algebra midterm, that'd be a big fat F. But on the math ACT, half the questions right is a 21, the average score for kids across the country. If you got 67% of the questions right, that'd be a D+ in school, but a nice 25 on the ACT, the average score for admission to great schools like Goucher and University of Vermont. And 87% correct, which is a B+ in school, is a beautiful 30 on the ACT, and about the average for kids who got into Tufts, U.C. Berkeley, University of Michigan, and Emory.

The math questions on the ACT are organized in order of difficulty, from easiest to hardest. In this book the drills that follow each skill are also arranged easiest to hardest. The easy questions are worth just as much as the hard ones. So don't rush and risk a careless error just to reach the hard questions. If an "easy" question seems difficult, take another look for what you are missing. Ask yourself, Which skill can I use? What is the easy way to do this question? After you complete this book, you will know!

You only need to get to the very hardest questions if you are shooting for 31+. However, since on the ACT you do **not** lose points for wrong answers, of course, you must put an answer for every question, even ones that you do not get to. Even if you are running out of time at question 40 out of 60, you must budget a few minutes to fill in an answer for the last 20 questions. It'd be crazy not to. Statistically, if you randomly fill in the last 20 ovals, you'll get 4 correct. That's worth about 2 points (out of 36) on your score! So keep an eye on the clock, and when there are a few minutes left, choose an answer for each remaining question.

About Brian Leaf

Six, maybe seven, people in the world know the ACT like Brian Leaf. Most are under surveillance in Iowa, and Brian alone is left to bring you this book.

Brian is the author of *McGraw-Hill's Top 50 Skills* SAT and ACT test-prep series. He is also the author of *Defining Twilight: Vocabulary Workbook for Unlocking the SAT, ACT, GED, and SSAT* (Wiley, 2009). Brian is Director of the New Leaf Learning Center in Western Massachusetts. He teaches ACT, SAT, and PSAT prep to thousands of students from throughout the United States. (For more information, visit his website www.BrianLeaf.com.) Brian also works with the Georgetown University Office of Undergraduate Admissions as an Alumni Interviewer, and is a certified yoga instructor and avid meditator. Read about Brian's yoga adventures in *Misadventures of a Garden State Yogi*.

Acknowledgments

Special thanks to all the students of New Leaf Learning Center for allowing me to find this book. Thanks to my agent, Linda Roghaar, and my Editor at McGraw-Hill, Anya Kozorez. Thanks to Pam Weber-Leaf for great editing tips, Larry Leaf for his one-liners, Zach Nelson for sage marketing advice, Ben Allison for his unparalleled math genius, Ian Curtis for assiduous editing, Corinne Andrews for designing the yoga sequence, Matthew Thompson for astute design help, Susan and Manny Leaf for everything, and of course, thanks most of all to Gwen and Noah for time, love, support, and in the case of Noah, Hexagons, Dexagons, and Shmexamons.

Pretest

This pretest contains questions that correspond to our 50 Skills. Take the test, and then check your answers with the answer key on page 11. The 50 Skills that follow contain more detailed solutions as well as instructions and drills for each type of question on the ACT.

DIRECTIONS: Solve each problem and choose the correct answer.
You are permitted to use a calculator on this test.

Note: Unless otherwise stated, all of the following should be assumed.

- Illustrative figures are NOT necessarily drawn to scale.
- Geometric figures lie in a plane.
- The word *line* indicates a straight line.
- The word *average* indicates arithmetic mean.

1 The lengths of the sides of a triangle are 3 consecutive even integers. If the perimeter of the triangle is 36 centimeters, what is the length, in centimeters, of the shortest side?

(A) 7
(B) 8
(C) 10
(D) 12
(E) 14

2 If $2x + 3 = 8x - 5$, then $x = ?$

(F) $\frac{4}{3}$
(G) $\frac{3}{4}$
(H) $\frac{1}{4}$
(I) $-\frac{1}{4}$
(K) -2

3 If $3x + 4y = 12$, what is y in terms of x?

(A) $3 - \frac{3}{4}x$
(B) $3 + \frac{3}{4}x$
(C) $4 - 3x$
(D) $12 - 3x$
(E) $x - 12$

4 So far, a student has earned the following scores on four 50-point quizzes during this marking period: 40, 37, 44, and 42. What score must the student earn on the fifth quiz to earn an average quiz score of 42 for the 5 quizzes?

(F) 42
(G) 45
(H) 47
(J) 49
(K) The student cannot earn an average of 42.

5 In the figure below, if $b = 100$, what is the value of c ?

  (A) 125
  (B) 130
  (C) 132
  (D) 143
  (E) 163

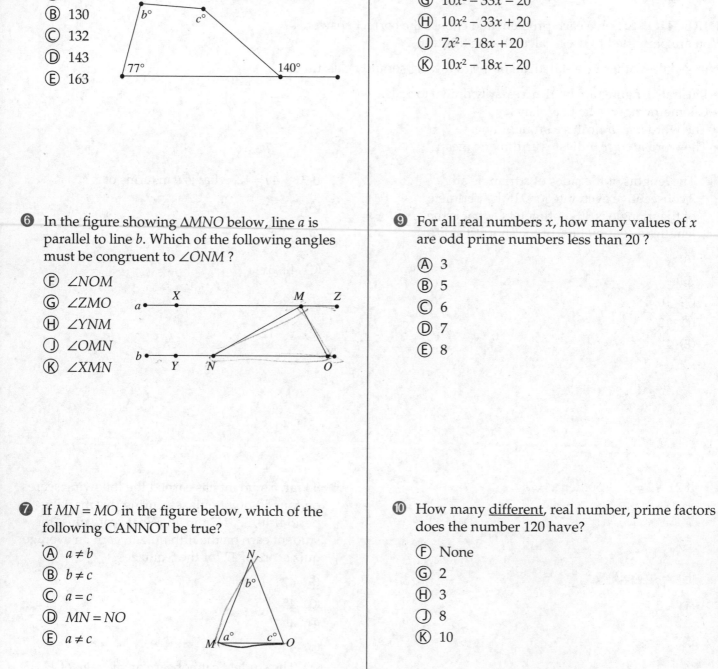

6 In the figure showing $\triangle MNO$ below, line a is parallel to line b. Which of the following angles must be congruent to $\angle ONM$?

  (F) $\angle NOM$
  (G) $\angle ZMO$
  (H) $\angle YNM$
  (J) $\angle OMN$
  (K) $\angle XMN$

7 If $MN = MO$ in the figure below, which of the following CANNOT be true?

  (A) $a \neq b$
  (B) $b \neq c$
  (C) $a = c$
  (D) $MN = NO$
  (E) $a \neq c$

8 For all x, $(2x - 5)(5x - 4) = ?$

  (F) $7x^2 + 20$
  (G) $10x^2 - 33x - 20$
  (H) $10x^2 - 33x + 20$
  (J) $7x^2 - 18x + 20$
  (K) $10x^2 - 18x - 20$

9 For all real numbers x, how many values of x are odd prime numbers less than 20 ?

  (A) 3
  (B) 5
  (C) 6
  (D) 7
  (E) 8

10 How many <u>different</u>, real number, prime factors does the number 120 have?

  (F) None
  (G) 2
  (H) 3
  (J) 8
  (K) 10

⑪ What is the least common multiple of 30, 40, and 50 ?

   Ⓐ 1200
   Ⓑ 800
   Ⓒ 600
   Ⓓ 400
   Ⓔ 50

⑫ What is the x intercept of the line that contains the points $(-2, 3)$ and $(4, 1)$ in the standard (x, y) coordinate plane?

   Ⓕ $(\frac{1}{2}, 0)$

   Ⓖ $(0, \frac{1}{2})$

   Ⓗ $(7, 0)$

   Ⓙ $(0, 7)$

   Ⓚ $(7, \frac{1}{2})$

⑬ If the slope of a line through the points $(-2, 5)$ and $(3, b)$ is -1, what is the value of b ?

   Ⓐ -2
   Ⓑ -1
   Ⓒ 0
   Ⓓ 1
   Ⓔ 2

⑭ In the standard (x, y) coordinate plane, what is the slope of the line with equation $3x - 5y = -12$?

   Ⓕ $\frac{12}{5}$

   Ⓖ 2

   Ⓗ $\frac{3}{5}$

   Ⓙ $-\frac{3}{5}$

   Ⓚ -2

⑮ The following chart shows the results when 31 high school juniors were asked to write the name of their favorite movie on a slip of paper and place it in a box.

Movie	Votes
Wedding Crashers	4
Across the Universe	6
Once	2
Lord of the Rings trilogy	5
Harry Potter movies	6
Superbad	3
Other	5

If a slip of paper is chosen at random from the box, which of the following is closest to the percent chance that the slip chosen will name *Wedding Crashers*?

   Ⓐ 4%
   Ⓑ 13%
   Ⓒ 31%
   Ⓓ 35%
   Ⓔ 69%

3

16 If $f(x) = -2x^2 - 3$, what is $f(-3)$?

(F) 33

(G) 15

(H) 3

(J) −21

(K) −39

17 A line in the standard (x, y) coordinate plane contains the points $P(3, -5)$ and $Q(-7, 9)$. What point is the midpoint of $\overline{PQ}$?

(A) (−2, 4)

(B) (−2, −4)

(C) (−2, 2)

(D) (−2, −2)

(E) (3, 3)

18 If $c = -2b - 2(2e - b)$, what happens to the value of c when b and e are both increased by 2 ?

(F) It is unchanged.

(G) It is increased by 2.

(H) It is increased by 4.

(J) It is decreased by 4.

(K) It is decreased by 8.

19 A square lot, with an area of 625 square feet, is completely fenced. What is the length, in feet, of the fence?

(A) 25

(B) 75

(C) 100

(D) 125

(E) 300

20 In the diagram below, if the radius of the large circle is 6 and the radius of the small circle is 4, what is the area of the shaded region?

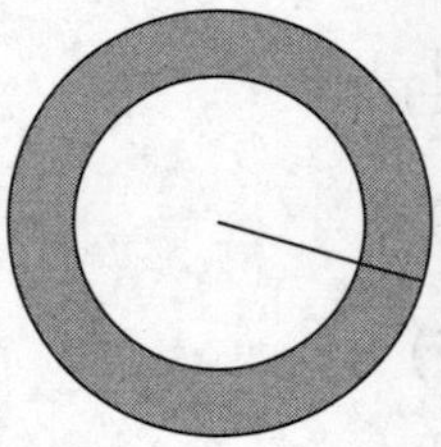

(F) 2

(G) 20

(H) 2π

(J) 20π

(K) $\sqrt{20}$

21 A triangle with a perimeter of 89 has one side that is 19 inches long. The lengths of the other two sides have a ratio of 3:7. What is the length, in inches, of the *longest* side of the triangle?

(A) 19

(B) 21

(C) 29

(D) 40

(E) 49

22 The number of students in three school clubs is shown in the following matrix.

$$\begin{bmatrix} \text{Yoga} & \text{Spanish} & \text{Best Buddies} \\ 20 & 18 & 36 \end{bmatrix}$$

The school paper reported that the student body of the school was comprised according to the following ratios.

$$\begin{bmatrix} 0.3 - \text{Freshmen} \\ 0.5 - \text{Sophomores \& Juniors} \\ 0.2 - \text{Seniors} \end{bmatrix}$$

Given these matrices, if participation in clubs is spread evenly among the grades, how many freshmen could be estimated to be in the Yoga Club?

(F) 0.2

(G) 0.3

(H) 20

(J) 10

(K) 6

23 Points A, B, C, D, and E are points on a line in that order. If B is the midpoint of $\overline{AC}$, C is the midpoint of $\overline{AD}$, and D is the midpoint of $\overline{AE}$, which of the following is the longest segment?

(A) $\overline{AB}$

(B) $\overline{AD}$

(C) $\overline{BC}$

(D) $\overline{BD}$

(E) $\overline{CE}$

24 In a right triangle, the measures of two sides are 6 and 10; which of the following could be the measure of the third side?

(F) 3

(G) 4

(H) 6

(J) 8

(K) 10

25 What is the shortest side of a triangle that is congruent to triangle PQR shown below?

(A) 3

(B) $3\sqrt{3}$

(C) 6

(D) $6\sqrt{3}$

(E) 12

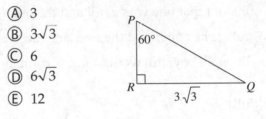

26 If triangle MNO (not shown) is similar to triangle PQR shown below, and has a shortest side with length 4.5, which of the following would be the measure of the longest side of triangle MNO?

(F) 4.5

(G) 6.5

(H) 9

(J) 12

(K) 15

27 If 5 percent of 20 percent of a number is 24 less than one-quarter of the number, what is the number?

Ⓐ 1
Ⓑ 5
Ⓒ 20
Ⓓ 50
Ⓔ 100

28 In Seth's refrigerator he found 2 jars of mustard. He estimated that one was $\frac{1}{3}$ full and the other was $\frac{2}{5}$ full. If he combined the two jars into one, approximately how full would the one combined jar be?

Ⓕ $\frac{1}{3}$ full
Ⓖ $\frac{3}{5}$ full
Ⓗ $\frac{2}{3}$ full
Ⓙ $\frac{9}{10}$ full
Ⓚ Completely full

29 If $x^2 - y^2 = 84$ and $x - y = 6$, what is the value of $x + y$?

Ⓐ 6
Ⓑ 8
Ⓒ 10
Ⓓ 12
Ⓔ 14

30 Which of the following expressions is equivalent to $(-2x^2y^2)^3$?

Ⓕ $-2x^5y^5$
Ⓖ $-8x^6y^6$
Ⓗ $2x^5y^5$
Ⓙ $8x^5y^5$
Ⓚ $8x^6y^6$

31 If $8m^2p^3 = m^5p$, what is m in terms of p?

Ⓐ $p^{2/3}$
Ⓑ $2p^{2/3}$
Ⓒ $8p^{2/3}$
Ⓓ $2p^2$
Ⓔ $8p^{-2}$

32. If Sawyer charges $20 per bottle of water and a flat fee of $25 even to discuss a sale, which of the following equations expresses Sawyer's total fees for x bottles of water?

F. $y = 45x$

G. $y = 25x + 20$

H. $y = 5x$

J. $y = 20x + 25$

K. $y = 45x$

33. Arthur Dent buys an ice cream sundae that contains one scoop of ice cream, one sauce, and either a cherry or pineapple wedge on top. He can choose chocolate, vanilla, strawberry, or banana ice cream; he can choose chocolate, caramel, or berry sauce; and he can choose either the cherry or the pineapple wedge for the top. How many different arrangements of these ingredients for Arthur's ice cream sundae are possible?

A. 9

B. 14

C. 24

D. 44

E. 64

34. For right triangle $\Delta\, XYZ$, shown below, what is tan Z ?

F. 0.2

G. 0.6

H. 0.75

J. 0.8

K. 0.9

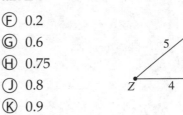

35. What is the value of θ, between 0 and 360, when $\sin \theta = -1$?

A. 0

B. 60

C. 135

D. 270

E. 330

36. Of the 18 socks in a drawer, 10 are solid blue, 4 are solid pink, and 4 are pink and blue. If Cherng-Mao randomly chooses a sock from the drawer, what is the probability that it will NOT be solid pink?

F. $\frac{1}{6}$

G. $\frac{2}{9}$

H. $\frac{4}{9}$

J. $\frac{5}{9}$

K. $\frac{7}{9}$

37 What are the values for x that satisfy the equation $(x + 4)(x - 3) = 0$?

(A) −4 and 4

(B) −3 and 3

(C) −12

(D) 4 and −3

(E) −4 and 3

38 If the graph of $y = ax^2 + bx + c$ is shown below, then the value of ac can be

 I. Positive

 II. Negative

 III. 0

(F) I only

(G) II only

(H) I or II

(J) I or III

(K) I, II, or III

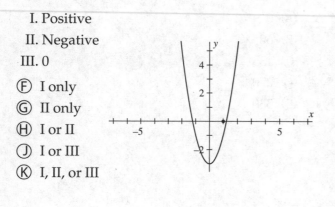

39 A circle in the standard (x, y) coordinate plane has center (4, 2) and radius of 5 coordinate units. Which of the following is an equation of the circle?

(A) $(x - 4)^2 - (y - 2)^2 = 5$

(B) $(x + 4)^2 + (y + 2)^2 = 5$

(C) $(x - 4)^2 + (y - 2)^2 = 5$

(D) $(x - 4)^2 - (y + 2)^2 = 25$

(E) $(x - 4)^2 + (y - 2)^2 = 25$

40 In the figure below, segment NO is a diameter of the circle, M is a point on the circle, and $MN = MO$. What is the degree measure of $\angle MNO$?

(F) 30

(G) 45

(H) 60

(J) 90

(K) Cannot be determined from the given information

41 Which of the following are solutions to $|n + 3| = 5$?

 I. 2

 II. −2

 III. −8

(A) I only

(B) III only

(C) II and III

(D) I and III

(E) I, II, and III

42 Which of the following statements is NOT true about the arithmetic sequence 20, 13, 6, −1, . . . ?

(F) The fifth term is −8.

(G) The sum of the first five terms is 30.

(H) The seventh term is −22.

(J) The common difference of terms is −7.

(K) The common ratio of consecutive terms is −7.

43 The temperature 357°F is the point at which mercury will boil. Since Fahrenheit and Celsius temperatures are related by the formula $F = \frac{9}{5}C + 32$, to the nearest degree, which of the following is the boiling point of mercury in degrees Celsius?

(A) 11°C

(B) 87°C

(C) 112°C

(D) 166°C

(E) 181°C

44 If $f(x) = 3x^2$, which of the following expresses $f(2p)$?

(F) $6p$

(G) $6p^2$

(H) $12p$

(J) $12p^2$

(K) $24p^3$

45 If a board 9 feet 10 inches long is cut in half, how long is each new piece?

(A) 4' 9"

(B) 4' 11"

(C) 5'

(D) 5' 2"

(E) 5' 5"

46 Suppose $0 < b < 1$. Which of the following has the greatest value?

(F) b^2

(G) b^3

(H) $\log b$

(J) $|b|$

(K) b^{-1}

47 What is the real value of x in the equation $\log_2 16 = \log_4 x$?

(A) 2

(B) 32

(C) 64

(D) 128

(E) 256

48 What is $(i - 2)(i - 3)$?

(F) $5 - 5i$

(G) $5 - 4i$

(H) $5 + i$

(J) 5

(K) -1

49 The table below shows the results of a survey in which 180 high school students voted for their favorite movie. Each student received one vote. According to the graph, how many more students favored *Superbad* than *Wedding Crashers*?

Movie	Votes	♥ = 20 votes
Superbad	♥ ♥ ♥ ♥ ♥	
The Dark Knight	♥ ❝	
Wedding Crashers	♥ ❝	
Godfather	♥	

(A) 3

(B) 3.5

(C) 35

(D) 70

(E) 79.5

50 Javier earned the following 7 test scores. What is the median?

85, 92, 82, 94, 90, 80, 79

(F) 80

(G) 82

(H) 85

(J) 86

(K) 90

Answer Key

1. C
2. F
3. A
4. H
5. D
6. K
7. B
8. H
9. D
10. H
11. C
12. H
13. C
14. H
15. B
16. J
17. A
18. K
19. C
20. J
21. E
22. K
23. E
24. J
25. A
26. H

27. E
28. H
29. E
30. G
31. B
32. J
33. C
34. H
35. D
36. K
37. E
38. G
39. E
40. G
41. D
42. K
43. E
44. J
45. B
46. K
47. E
48. F
49. D
50. H

Use the Answers

For 11 years you've been trained to solve math problems the long way. "No shortcuts!" Mrs. Nicholas always said. Mrs. Nicholas was the best math teacher that I ever had, and her advice was correct for math class. But the ACT is multiple-choice, and we can save time and energy by using the answers; simply test the answer choices to see which one works. To make this even easier, when there are fractions, π, or square roots in the questions or answers, you can convert them to decimals. "Use the Answers" works best when there are **variables or unknowns** in the question and **numbers** in the answer choices.

Let's try this on the question from the Pretest:

1. The lengths of the sides of a triangle are 3 consecutive even integers. If the perimeter of the triangle is 36 centimeters, what is the length, in centimeters, of the shortest side?

   A. 7   **B.** 8   **C.** 10   **D.** 12   **E.** 14

Solution: This question looks pretty tough to many students, but "Use the Answers" makes it easy! You can answer this question by setting up an algebraic equation, or you can just "Use the Answers." We want 3 **consecutive even integers** that add up to 36. That means we want 3 even numbers in a row, like 2, 4, and 6. Try each choice as the shortest side and see which one makes the situation work:

Ⓐ 7 is not even.
Ⓑ $8 + 10 + 12 \neq 36$
Ⓒ $10 + 12 + 14 = 36$
Ⓓ $12 + 14 + 16 \neq 36$
Ⓔ $14 + 16 + 18 \neq 36$

Choice C is correct since $10 + 12 + 14 = 36$. *Careless error buster:* Notice that the question asks for the shortest, not the longest, side.

Correct answer: C.

ACT Math Mantra #1
When you see <u>variables</u> or <u>unknowns</u> in the question and <u>numbers</u> in the answer choices, "Use the Answers." Convert fractions, π, or √ to decimals.

Use the Answers Drills

Easy

1 If $3x - 4 = -10$, then $x = ?$

- (A) -2
- (B) -1
- (C) 0
- (D) 2
- (E) 5

Medium

2 $\sqrt{40} + \sqrt{160} = ?$

- (F) $6\sqrt{10}$
- (G) $7\sqrt{10}$
- (H) $10\sqrt{2} + 10\sqrt{4}$
- (J) $20\sqrt{10}$
- (K) $\sqrt{200}$

3 If $n - \frac{2}{3} = \frac{11}{27}$, then $n = ?$

- (A) -1
- (B) $-\frac{7}{27}$
- (C) 0
- (D) $\frac{29}{27}$
- (E) $\frac{9}{30}$

4 What is the largest integer value of p that satisfies the inequality $\frac{15}{18} \geq \frac{p}{12}$?

- (F) 8
- (G) 9
- (H) 10
- (J) 11
- (K) 12

Hard

5 Jaleesa earned a score of 160 on a recent 50-question spelling contest. The scoring for the contest was +4 for each correct answer, −1 for each incorrect answer, and 0 for each unanswered question. If she answered every question, what is the maximum number of questions that Jaleesa could have answered correctly?

- (A) 8
- (B) 20
- (C) 40
- (D) 42
- (E) 46

Super Easy Algebra

Every ACT has several easy algebra questions. The key on these is to overcome any algebra phobia still lingering from seventh grade. You're older and smarter than you were then, and this stuff is predictable. It follows rules. As long as you stay present (see Yoga, page 66), you'll get them right. If they give you trouble, do and redo the Pretest question and the drills until you could teach them. Then do that—take a friend and give a lesson. You have my permission to throw out a few detentions if needed.

Let's take a look at the Pretest:

2. If $2x + 3 = 8x - 5$, then $x = ?$

F. $\frac{4}{3}$ **G.** $\frac{3}{4}$ **H.** $\frac{1}{4}$ **J.** $-\frac{1}{4}$ **K.** -2

Solution: Just use algebra to get x alone.

$$2x + 3 = 8x - 5$$
$$\underline{-2x \qquad -2x}$$
$$3 = 6x - 5 \qquad \text{Subtract } 2x \text{ from both sides.}$$
$$\underline{+5 \qquad +5}$$
$$8 = 6x \qquad \text{Add 5 to both sides.}$$
$$\underline{\div 6 \quad \div 6} \qquad \text{Divide both sides by 6.}$$
$$\frac{8}{6} = \frac{4}{3} = 1.33 = x$$

Correct answer: F.

See, easy. But if after doing the drills and reviewing the solutions, you still don't agree, you could just "Use the Answers" to solve any of these algebra questions!

ACT Math Mantra #2
When you see "then $x = ?$"
complete the algebra or just "Use the Answers."

Super Easy Algebra Drills

Easy

1 If $6x + 6 = 36 + 3x$, then $x = ?$

Ⓐ 3

Ⓑ 9

Ⓒ 10

Ⓓ 33

Ⓔ $\frac{9}{5}$

Medium

2 If $3(x + 4) + 2x = 2(x - 2) - 4$, then $x = ?$

Ⓕ $-\frac{20}{5}$

Ⓖ $-\frac{20}{3}$

Ⓗ $-\frac{10}{3}$

Ⓙ 2

Ⓚ $\frac{11}{5}$

3 In a certain situation, the maximum height H, in meters, that a ball bounces is given by the equation $H = \frac{3}{5}F - 0.02$, where F is the force, in newtons, with which the ball is bounced. What force, in newtons, must be applied for the height to be 0.16 meters?

Ⓐ 0.1

Ⓑ 0.2

Ⓒ 0.3

Ⓓ 0.4

Ⓔ 0.5

Hard

4 Which of the following is the set of all real numbers x such that $x - 2 < x - 4$?

Ⓕ The set containing only zero.

Ⓖ The set containing all positive real numbers.

Ⓗ The set containing all negative real numbers.

Ⓙ The set containing all real numbers.

Ⓚ The empty set.

Super Easy Algebra— Algebraic Manipulation

The first type of Super Easy Algebra questions (Skill 2) gave you an equation and asked you to solve for x. The answers to these questions were always numbers. The questions in this Skill are similar, but the answer choices will have variables as well as numbers in them. Some teachers call these algebraic manipulation questions. I don't like that term; it sounds brutal, like a heinous, medieval torture. And, on the contrary, solving an algebraic equation is actually fun and easy—well maybe just easy.

Some of these questions ask, What is y in terms of x and z? For some reason kids seem to hate these. They come into my office and say, "This question says, 'What is y in terms of x and z' and I have no idea what that means, so I skipped it." I love these though, because they are so easy to gain points on. "What is y in terms of x and z" is just a fancy way of saying "solve for y" or "use algebra to get y alone." It's no different than "$y = ?$" Whichever letter is after the "what is . . ." is the variable that you solve for.

Let's take a look at the Pretest:

3. If $3x + 4y = 12$, what is y in terms of x?

    A. $3 - \dfrac{3}{4}x$ **B.** $3 + \dfrac{3}{4}x$ **C.** $4 - 3x$ **D.** $12 - 3x$ **E.** $x - 12$

Solution: Just use algebra to get y alone.

$$3x + 4y = 12$$
$$\underline{-3x \qquad\quad -3x}$$
$$4y = 12 - 3x$$

Subtract $3x$ from both sides.
Divide both sides by 4.

$$\boxed{y = 3 - \dfrac{3}{4}x}$$

Correct answer: A

ACT Math Mantra #3
"What is y in terms of x and z" is just a fancy way of saying "solve for y"
or "use algebra to get y alone."

Super Easy Algebra—Algebraic Manipulation Drills

Easy

❶ If $I = PRT$, then which of the following is an expression for R in terms of I, P, and T ?

  Ⓐ IPT

  Ⓑ $I - PT$

  Ⓒ $\dfrac{I}{PT}$

  Ⓓ $\dfrac{PT}{I}$

  Ⓔ $\dfrac{2}{IPT}$

Medium

❷ For all pairs of real numbers N and P where $N = 5P + 4$, $P = ?$

  Ⓕ $\dfrac{N}{5} - 4$

  Ⓖ $\dfrac{N}{5} + 4$

  Ⓗ $5N - 4$

  Ⓙ $\dfrac{N - 4}{5}$

  Ⓚ $\dfrac{N + 4}{5}$

Hard

❸ If $x = 3m - 4$ and $y = 6 - m$, which of the following expresses y in terms of x ?

  Ⓐ $y = 2x - 4$

  Ⓑ $y = \dfrac{x + 10}{3}$

  Ⓒ $y = \dfrac{2 + x}{3}$

  Ⓓ $y = x - \dfrac{4}{3}$

  Ⓔ $y = \dfrac{14 - x}{3}$

"Mean" Means Average

Every ACT has a "mean" question. If you didn't know the term before and you do by the end of this page, you will gain points! "Mean" is just a fancy term for "average." So when a question asks for either the "mean" or the "average" of a bunch of numbers, just add them up and divide by how many numbers there are:

$$\text{Average} = \frac{\text{sum}}{\text{number of items}}$$

The ACT throws only two "mean" curveballs. First, instead of giving you a list of numbers, they'll show you the data in a table or graph. This throws kids; it surprises them and they don't know what to do. But since you know to expect it, it's easy. Just identify the data and use our average formula: add up the numbers and divide by how many there are. Once you drill this topic, you'll never miss a "mean" or "average" question again!

The second curveball is that sometimes the ACT whips out a question about median or mode. These are usually rated "medium" or "hard," only because most kids don't know what these terms mean! **Median** is the middle number in a list of numbers, and **mode** is the number that occurs the most often. Consider this list: 5, 7, 9, 9, 11, 12, 12, 15, 15, 15, 15. The median is 12 (it's in the middle of the list). The mode is 15 (it occurs most often, 4 times). When you see a median or mode question, just rewrite the data as a list in order.

Let's look at the question from the Pretest:

4. So far, a student has earned the following scores on four 50-point quizzes during this marking period: 40, 37, 44, and 42. What score must the student earn on the fifth quiz to earn an average quiz score of 42 for the 5 quizzes?

   F. 42   **G.** 45   **H.** 47   **J.** 49
   K. The student cannot earn an average of 42.

Solution: You can set up the average formula and solve for the fifth quiz grade; or you can just "Use the Answers," trying each answer choice and seeing which one works. Try both methods and see which one feels more comfortable for you.

$$\text{Average} = \frac{\text{sum}}{\text{number of item:}} \quad \Rightarrow \quad 42 = \frac{40 + 37 + 44 + 42 + x}{5}$$

$210 = 40 + 37 + 44 + 42 + x$ Multiply both sides by 5.
$210 = 163 + x$ Add the numbers.
$47 = x$ Subtract 163 from both sides.

Correct answer: H

ACT Math Mantra #4
When you see the word "mean" or "average" on the ACT, use $\text{Average} = \dfrac{\text{sum}}{\text{number of items}}$.

"Mean" Means Average Drills

Easy

① Abida wants to find the average cost of a pint of ice cream at a local store. If the five different pints that the store sells are priced at $1.80, $2.10, $1.90, $2.80, and $3.20, what is the average price?

Ⓐ $2.20
Ⓑ $2.24
Ⓒ $2.32
Ⓓ $2.36
Ⓔ $2.42

② During the summer Terrell mows lawns. So far this summer, he earned $230 the first week, $250 the second week, and $257 the third week. His mean earnings per week for the three-week period are closest to which of the following?

Ⓕ $230
Ⓖ $235
Ⓗ $240
Ⓙ $245
Ⓚ $250

Medium

③ In a bowling league match, Jimbo has bowled scores of 180 and 210 so far. He will bowl one more game. If he wants his average for the three games to be at least 205, what is the minimum score he needs for the third game?

Ⓐ 185
Ⓑ 205
Ⓒ 215
Ⓓ 225
Ⓔ 245

④ According to the chart below, what is the average number of students enrolled per section in World History?

Ⓕ 22
Ⓖ 21
Ⓗ 22
Ⓙ 23
Ⓚ 25

Course	Section	Enrollment
World History	A	19
	B	25
	C	19
U.S. History	A	31

⑤ The median of a set of 9 consecutive integers is 42. What is the greatest of these 9 numbers?

Ⓐ 9
Ⓑ 43
Ⓒ 44
Ⓓ 45
Ⓔ 46

The Six-Minute Abs of Geometry: Angles

Every ACT has two or three geometry questions that ask you to find the measure of an angle. Many students panic, "I don't remember the million postulates we learned." Good news, you only need a small handful of those million postulates to solve most geometry questions.

Skills 5 to 7 teach you everything that you need to find the measure of an angle on the ACT; memorize these skills and you will gain points, guaranteed.

❶ Vertical angles are equal.

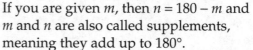

❷ The angles in a linear pair add up to 180°.

$$m + n = 180°$$

If you are given m, then $n = 180 - m$ and m and n are also called supplements, meaning they add up to 180°.

❸ The angles in a triangle add up to 180°.
The angles in a 4-sided shape add up to 360°.
The angles in a 5-sided shape add up to 540°.
The angles in a 6-sided shape add up to 720°.

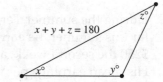

$$x + y + z = 180$$

Now, let's look at the question from the Pretest:

5. In the figure below, if $b = 100$, what is the value of c?

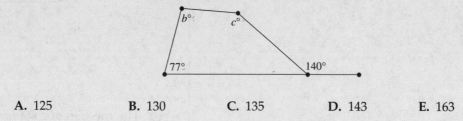

A. 125   **B.** 130   **C.** 135   **D.** 143   **E.** 163

Solution: First, always mark all info from the question into the diagram. This helps you see what to do next. This question uses two of our strategies: linear pair and 360° in a four-sided shape. Whenever you are given the measure of one angle in a linear pair of angles, determine the measure of the other angle in the pair. This always brings you toward the right answer; it's what the ACT wants you to do. So the angle next to the 140° must be $180 - 140 = 40°$. Now, we have three of four angles which must add up to 360°, so the fourth must be $360 - 100 - 40 - 77 = 143°$.

Correct answer: D.

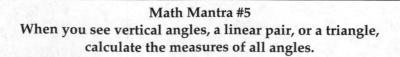

Math Mantra #5
When you see vertical angles, a linear pair, or a triangle, calculate the measures of all angles.

The Six-Minute Abs of Geometry: Angles Drills

Easy

① If the measure of an angle is 42.5°, what is the measure of its supplement?

Ⓐ 47.5°
Ⓑ 57.5°
Ⓒ 132.5°
Ⓓ 137.5°
Ⓔ Cannot be determined from the given information

② In the figure below, if $y = 45$ and points A, B, and C lie on the same line, what is the value of z ?

Ⓕ 204
Ⓖ 135
Ⓗ 84
Ⓙ 72
Ⓚ 69

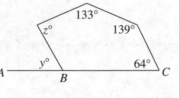

Medium

③ In the figure below, with angles as marked, if $x = 60°$, what is the value of y ?

Ⓐ 50
Ⓑ 60
Ⓒ 70
Ⓓ 90
Ⓔ 110

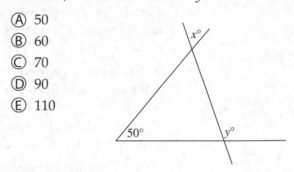

Hard

④ In the figure below, $ABCD$ is a trapezoid, E lies on $\overrightarrow{AD}$, and angle measures are as indicated. If the measure of angle $BAD = 49°$, what is the measure of $\angle DBA$?

Ⓕ 110
Ⓖ 112
Ⓗ 131
Ⓙ 156
Ⓚ 189

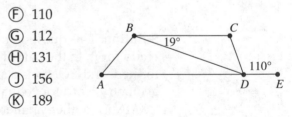

The Six-Minute Abs of Geometry: Parallel Lines

Most ACTs have one question involving parallel lines. In math class, learning about parallel lines might have seemed pretty tricky—alternate interior angles, corresponding angles, same-side interior angles. . . . We don't need all that vocab for the ACT. We just need to know:

• Parallel lines are two lines that never touch.
• If two parallel lines are crossed by another line (called a transversal), then eight angles form.
• These eight angles are of two types, big or little. All bigs are equal, and all littles are equal.

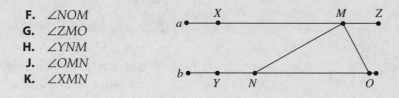

This is enough to answer any parallel-lines ACT question.

Here's the question from the Pretest:

6. In the figure showing ∆MNO below, line *a* is parallel to line *b*. Which of the following angles must be congruent to ∠ONM ?

    F. ∠NOM
    G. ∠ZMO
    H. ∠YNM
    J. ∠OMN
    K. ∠XMN

Solution: This question may look tough at first, but it uses Skill 6 exactly, and we're ready for it. In the pair of parallel lines crossed by segment *NM*, several angles are formed, and ∠ONM is a smaller-looking angle and is congruent (*congruent* is just a fancy geometry word for "equal") to all other smaller-looking angles of the angles formed. ∠XMN is another smaller-looking angle, so ∠ONM = ∠XMN. ∠YNM and ∠ZMN are bigger-looking angles and are equal to each other, but not to ∠ONM.

Correct answer: K.

ACT Math Mantra #6
When you see two parallel lines that are crossed by another line,
eight angles are formed, and all the bigger-looking angles are equal,
and all the smaller-looking angles are equal.

The Six-Minute Abs of Geometry: Parallel Lines Drills

Easy

1 In the figure below, $m \parallel n$. If $y = 37.5°$, what is the value of x?

Ⓐ 42.5
Ⓑ 100
Ⓒ 137.5
Ⓓ 142.5
Ⓔ 180

2 If $x = 73°$ and $p \parallel q$ and $m \parallel n$ in the four lines shown, what is the value of z?

Ⓕ 17
Ⓖ 73
Ⓗ 90
Ⓙ 97
Ⓚ 107

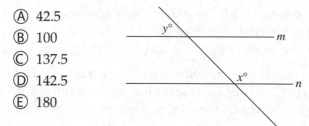

Medium

3 In the figure below, $m \parallel n$. If $z = 42.5$ and $x = 125$, then $y = ?$

Ⓐ 97.5
Ⓑ 87.5
Ⓒ 57.5
Ⓓ 47.5
Ⓔ 27.5

4 If $z = 55°$ and $y = 75°$ in the picture below, then which of the following must be true?

Ⓕ $p \parallel q$
Ⓖ $x + y = 180°$
Ⓗ $z = x$
Ⓙ $x + z = 180°$
Ⓚ $y = 2x$

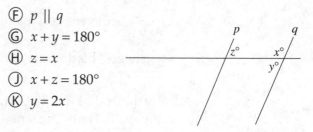

Hard

5 In the figure below, V, W, X, and Y are collinear; Z, W, and T are collinear; and the angles at V, X, and T are right angles, as marked. Which of the following statements is NOT justifiable from the given information?

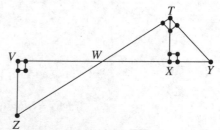

Ⓐ $\overline{YT}$ is perpendicular to $\overline{ZT}$.
Ⓑ $\angle VZW$ is congruent to $\angle XTW$.
Ⓒ $\overline{XY}$ is congruent to $\overline{TX}$.
Ⓓ $\triangle ZVW$ is similar to $\triangle TXW$.
Ⓔ $\overrightarrow{VZ}$ is parallel to $\overleftrightarrow{TX}$.

The Six-Minute Abs of Geometry: Triangles

Sir Bedevere: . . . And that, my liege, is how we know the Earth to be banana-shaped.

King Arthur: This new learning amazes me, Sir Bedevere. Explain again how sheeps' bladders may be employed to prevent earthquakes.

Monty Python and the Holy Grail (20th Century Fox, 1975)

By the end of this Skill, you will have learned all that you need to know about angles for the ACT. That's great news because **every** ACT includes several angle questions, and now you can always get them right. You know what to expect, you know what to use, and you will earn more points!

Back in geometry class, you had a full chapter with 14 theorems classifying triangles. Here are two that matter for the ACT.

❶ If a triangle is isosceles (a fancy term for having two equal sides), then the two angles opposite the equal sides are also equal.

❷ If a triangle is equilateral (a fancy term for having all sides equal), then it has all equal angles of 60° each.

Let's take a look at the question from the Pretest:

7. If $MN = MO$ in the figure below, which of the following CANNOT be true?

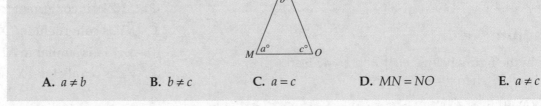

A. $a \neq b$ **B.** $b \neq c$ **C.** $a = c$ **D.** $MN = NO$ **E.** $a \neq c$

Solution: As soon as you are given info in the question, mark it in the diagram. This will remind you which geometry Skill to use: since two sides are equal in the triangle, the two angles opposite the two sides are also equal. So $b = c$. Therefore, $b \neq c$ definitely CANNOT be true. All other answers are possible since segment NO might also equal segments MN and MO.

Correct answer: B

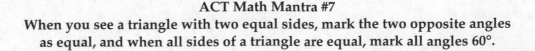

ACT Math Mantra #7
When you see a triangle with two equal sides, mark the two opposite angles as equal, and when all sides of a triangle are equal, mark all angles 60°.

The Six-Minute Abs of Geometry: Triangles Drills

Easy

① If triangle *MON*, shown below, is equilateral, what is the value of *x* ?

Ⓐ 30
Ⓑ 60
Ⓒ 90
Ⓓ 120
Ⓔ 150

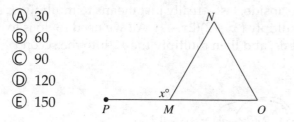

Medium

② In the figure below, $\overline{PQ}$ is parallel to $\overline{RT}$ with *A* on $\overline{PQ}$ and *S* on $\overline{RT}$. Also *PS = SQ*, and the measure of ∠*APS* is 108°. What is the measure of ∠*PSQ* ?

Ⓕ 36°
Ⓖ 48°
Ⓗ 72°
Ⓙ 108°
Ⓚ 180°

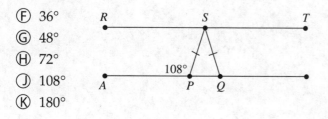

Hard

③ In △*ABC*, $\overline{AB} = \overline{AC}$ and the measure of ∠*C* is 28°. What is the measure of ∠*A* ?

Ⓐ 28°
Ⓑ 52°
Ⓒ 62°
Ⓓ 94°
Ⓔ 124°

④ If the area of a right triangle is 72, and the measure of one leg is 12, which of the following could be the value of one of its angles?

Ⓕ 15°
Ⓖ 30°
Ⓗ 45°
Ⓙ 72°
Ⓚ 144°

FOIL

If you're uncomfortable with the word *FOIL,* if you've grown to fear or detest it, then relax and know that FOIL—first, outside, inside, last—really just means to multiply. It's totally no biggie. Let's say we want to multiply $(3x + 2)(2x - 4)$. All we need to do is to multiply the $3x$ into the second parenthesis, and then multiply the $+2$ into the second parenthesis. That's FOIL. Watch:

First, multiply $3x$ by $(2x - 4)$. $(3x + 2)(2x - 4) = 6x^2 - 12x$

Then multiply 2 by $(2x - 4)$. $(3x + 2)(2x - 4) = 4x - 8$

And collect like terms: $6x^2 - 12x + 4x - 8 = \boxed{6x^2 - 8x - 8}$

Let's try this out on the question from the Pretest.

8. For all x, $(2x - 5)(5x - 4) = ?$

   F. $7x^2 + 20$
   G. $10x^2 - 33x - 20$
   H. $10x^2 - 33x + 20$
   J. $7x^2 - 18x + 20$
   K. $10x^2 - 18x - 20$

Solution: $(2x - 5)(5x - 4)$ means multiply "$2x$" into the second parenthesis, multiply "-5" into the second parenthesis, and then collect like terms.

$2x(5x - 4) = 10x^2 - 8x$ Multiply "$2x$" into the second parenthesis.
$-5(5x - 4) = -25x + 20$ Multiply "-5" into the second parenthesis.
$10x^2 - 8x + (-25x) + 20$
$10x^2 - 33x + 20$ Collect like terms: $-8x - 25x = -33x$

Correct answer: H

ACT Math Mantra #8

When you see an expression like $(2x - 5)(5x - 4)$, multiply the first number and letter into the second parenthesis, multiply the second number and letter into the second parenthesis, and then collect matching terms.

FOIL Drills

Easy

❶ Which of the following expressions is equivalent to $(2x - 2)(x + 5)$?

Ⓐ $2x^2 + 12x - 10$
Ⓑ $2x^2 + 12x + 10$
Ⓒ $2x^2 - 12x + 10$
Ⓓ $2x^2 + 8x - 10$
Ⓔ $2x^2 - 8x + 10$

❷ The expression $(3m - 2)(5m + 3)$ is equivalent to

Ⓕ $15m^2 - 6$
Ⓖ $15m^2 - m - 6$
Ⓗ $8m^2 - 6$
Ⓙ $8m^2 + 3m - 6$
Ⓚ $8m^2 - m - 6$

Medium

❸ The length of a side of a square is represented as $(4x + 3)$ inches. Which of the following general expressions represents the area of the square, in square inches?

Ⓐ $16x^2 + 24x + 9$
Ⓑ $16x^2 - 24x + 6$
Ⓒ $16x^2 - 12x + 9$
Ⓓ $16x^2 + 9$
Ⓔ $8x + 6$

❹ What is the coefficient of x^7 in the product of the two polynomials below?

$$(x^4 - 3x^3 + 6x^2 - 2x + 1)(4x^3 - 5x^2 + 2x)$$

Ⓕ 0
Ⓖ 2
Ⓗ 4
Ⓙ 7
Ⓚ 9

Hard

❺ A rectangular room that is 1 foot longer than it is wide has an area of 90 square feet. How many feet long is the room?

Ⓐ 8
Ⓑ 9
Ⓒ 10
Ⓓ 12
Ⓔ 15

Students come into my office and say, "I can't do this question because I don't know what *consecutive* means." I get to say, "Oh cool, it just means numbers in a row, like 4, 5, 6." Then the question becomes easy. In fact, for some questions, the only hard part is knowing the vocabulary, and once you know the terms, the questions are easy. That's why I love these next four Skills; memorize the terms, practice using them in the drills, and you will absolutely gain points, guaranteed. Also, once you know the terms, watch for them and <u>underline</u> them when they appear. That will eliminate many careless errors.

Here are the first nine math vocabulary terms:

Real number—You can ignore this term, it just means any number: $-3, -2.2, 0, \sqrt{2}, \pi$.

Constant term—This word really throws some kids, but it just means a letter in place of number, kinda like a variable, except that it won't vary.

Integer—Numbers without decimals or fractions: $-3, -2, -1, 0, 1, 2, 3, \ldots$

Even/odd—Even numbers: $2, 4, 6, 8, \ldots$; odd numbers: $1, 3, 5, 7, \ldots$
　　　　The number 0 is considered even.

Positive/negative—Positive numbers are greater than 0.
　　　　　　Negative numbers are less than 0.

Consecutive numbers—Numbers in a row: 7, 8, 9, 10.

Different numbers—Numbers that are . . . ummm . . . different.

Prime—A number whose only factors are 1 and itself. The numbers $2, 3, 5, 7, 11, 13, 17, \ldots$ are prime.
　　　Note: The number 1 is NOT considered prime, and the number 2 is the *only* even prime number.

Units digit—Just a fancy term for the "ones" digit in a number, like the 2 in 672.

Let's take a look at the question from the Pretest:

9. For all real numbers x, how many values of x are odd prime numbers less than 20 ?

　　A. 3　　　**B.** 5　　　**C.** 6　　　**D.** 7　　　**E.** 8

Solution: List the primes less than 20: 2, 3, 5, 7, 11, 13, 17, 19. All are odd except 2. Remember that 1 is not a prime number. So there are 7 odd prime numbers less than 20.

Correct answer: D

ACT Math Mantra #9
Anytime you see a math vocab term, underline it.

Math Vocab Drills

Easy

❶ For all real numbers x, what integer values satisfy the equation $2x^2 = 50$?

Ⓐ 25 only

Ⓑ 5 only

Ⓒ −5 only

Ⓓ −5 and 5 only

Ⓔ −5, 5, and 25 only

❷ If the sum of three consecutive odd integers is 33, what is the value of the largest of the three integers?

Ⓕ 7

Ⓖ 9

Ⓗ 11

Ⓙ 13

Ⓚ 15

Medium

❸ If 2 consecutive prime numbers have a sum of 42, what is the larger number?

Ⓐ 17

Ⓑ 19

Ⓒ 23

Ⓓ 27

Ⓔ 29

❹ Two consecutive odd integers have a sum of −32. Which of the following is NOT a justified conclusion?

Ⓕ Both numbers are negative.

Ⓖ The lesser number is −15.

Ⓗ Both numbers are greater than −20.

Ⓙ Both numbers are less than −10.

Ⓚ One of the numbers is a factor of 60.

Hard

❺ Which of the following is a list of all positive integers between 40 and 46 that are not consecutive odd prime numbers?

Ⓐ 42, 44, and 45

Ⓑ 42, 43, and 44

Ⓒ 41, 42, 43, and 44

Ⓓ 41 and 45

Ⓔ 42 and 44

More Math Vocab

Here are three more math vocab terms. Memorize them, practice using them, and remember to underline them in questions. That will avoid heaps of careless errors.

Factors—Numbers that divide into a number evenly (i.e., without a remainder).
Example: The factors of 48 are 1, 2, 3, 4, 6, 8, 12, 16, 24, 48.

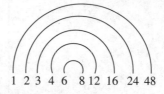

1 2 3 4 6   8 12 16   24 48

When asked for the factors of a number, make a list of pairs like the one shown above. This eliminates the possibility of missing any.

Greatest common factor—The largest factor shared by several given numbers.
Example: The greatest common factor of 48 and 32 is 16. The number 16 is the largest number that is a factor of both 48 and 32.

Prime factors—The factors of a number that are also prime numbers.
(Remember, a prime number is a number whose only factors are 1 and itself.)
Example: The prime factors of 48 are 2 and 3. These are the factors of 48 that also happen to be prime numbers.

Now, let's take a look at the question from the Pretest:

10. How many <u>different</u>, real number, prime factors does the number 120 have?

F. None   **G.** 2   **H.** 3   **J.** 8   **K.** 10

Solution: To complete this question, we factor 120 just as we factored 48 above.

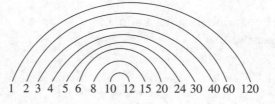

1   2 3 4 5 6 8 10 12 15 20 24 30 40 60 120

Then we circle any prime numbers in the list. The numbers 2, 3, and 5 are the only prime numbers in the list, so there are 3 prime factors.

Correct answer: H.

ACT Math Mantra #10
Don't be intimidated by fancy vocabulary terms.

More Math Vocab Drills

Easy

1 Which of the following lists the factors of 60 ?

(A) 1, 10, 20, 30, and 60

(B) 2, 4, 6, 10, 12, 20, 30, and 60

(C) 1, 2, 3, 4, 5, 6, 10, 12, 15, 20, 30, and 60

(D) 1, 2, 3, 4, 5, 6, and 10

(E) 60, 120, 180, and 240

2 Which of the following is an odd number that is a factor of 140 ?

(F) 2

(G) 13

(H) 35

(J) 37

(K) 51

Medium

3 How many prime integer factors does the number 50 have?

(A) None

(B) 1

(C) 2

(D) 6

(E) 10

Hard

4 If a and b are positive integers greater than 1, such that the greatest common factor of a^3b and a^2b^2 is 36, then which of the following could a equal?

(F) 36

(G) 18

(H) 9

(J) 6

(K) 3

5 If a, b, and c are different odd prime numbers, how many factors does $2abc$ have?

(A) 4

(B) 5

(C) 15

(D) 16

(E) 24

Multiples Vocab

Almost every ACT has a "multiples" question. We know the recipe for the ACT, one multiples question, one average question, etc. These 50 Skills review exactly what you need. Multiples are great to review; you knew this inside-out back in 5th grade. Mrs. Shortino taught it, you did the 7 minutes of homework, and then went straight to the kickball field. Now, you haven't thought about multiples for 6 years. So, here's a quick review, and guess what, now that you're older, anything that seemed hard in 5th grade will be easy!

Multiples—All the numbers that are divisible by a certain number.
   Example: The multiples of 3 are 3, 6, 9, 12, 15, 18, 21, etc.

Least common multiple—The lowest number that is a multiple of several numbers.
   Example: What is the least common multiple of 10, 15, and 20 ?

The trick to finding the least common multiple is to list the multiples of the largest given number and then choose the lowest one that is also a multiple of the other given numbers. This is a great trick; we start with the biggest number because it has fewest multiples to consider and will save time and energy.

So to find the least common multiple of 10, 15, and 20, list the multiples of 20: 20, 40, 60, 80, 100. And choose the first one that is also a multiple for both 10 and 15. The number 60 works because it is the lowest number on the list that is a multiple of 10, 15, and 20.

Of course, for most questions, you'll also just be able to "Use the Answers," which will probably be even faster. *Note:* You can use this same strategy when an ACT question asks for the **lowest common denominator**; it's basically the same thing.

Now, let's take a look at the question from the Pretest:

11. What is the least common multiple of 30, 40, and 50 ?

   A. 1200   **B.** 800   **C.** 600   **D.** 400   **E.** 50

Solution: This question could be difficult, but "Use the Answers" makes it easy! Just test each answer choice by dividing it by 30, 40, and 50. The number 600 is the lowest number on the list that is divisible by 30, 40, and 50. Notice that 1200 works also, but 600 is the **lowest**. Make sure to try all choices.

Correct answer: C

> **ACT Math Mantra #11**
> **To find the "least common multiple" or "lowest common denominator,"**
> **"Use the Answers!"**

Multiples Vocab Drills

Easy

1 What is the lowest number that is a multiple of 10, 12, and 15 ?

Ⓐ 1800
Ⓑ 900
Ⓒ 120
Ⓓ 60
Ⓔ 30

Medium

2 126 is the least common multiple for which pair of numbers?

Ⓕ 2 and 62
Ⓖ 9 and 14
Ⓗ 6 and 22
Ⓙ 3 and 42
Ⓚ 5 and 25

3 What is the least common multiple of 60, 50, and 80 ?

Ⓐ 60
Ⓑ 120
Ⓒ 1002
Ⓓ 1200
Ⓔ 12,000

4 If m and p are positive integers, what is the least common multiple of 4, $3m$, $5p$, and $6mp$?

Ⓕ $30mp$
Ⓖ $60mp$
Ⓗ $60mp^2$
Ⓙ $120mp$
Ⓚ $120m^2p$

Hard

5 A bag of peanuts could be divided among 8 children, 9 children, or 10 children with each getting the same number, and with 2 peanuts left over in each case. What is the smallest number of peanuts that could be in the bag?

Ⓐ 722
Ⓑ 362
Ⓒ 182
Ⓓ 56
Ⓔ 29

Brian's Math Magic Trick
#1 - Multiples
Answer the following questions aloud.

What is 2×2 ?
What is 4×2 ?
What is 8×2 ?
What is 16×2 ?
What is 32×2 ?
What is 64×2 ?

Name a vegetable.

Turn to the solutions page to be amazed.

Fancy Graphing Vocab

Kids come into my office and say, "I skipped number 10 because I never learned what an 'ordered pair' is." My job is easy. I get to say, "Yea, that's a funny thing, 'ordered pair' is just a fancy term for a normal (x, y) point on a graph that you are used to; it's just a silly name. Don't be intimidated by it!" I love teaching vocabulary; know the terms, and you'll get the question correct!

So, here are five fancy graphing terms:

Standard (x, y) coordinate plane—A fancy term for the normal grid that you graph lines on.

Cartesian plane—Another fancy term for the normal grid that you graph lines on.

Ordered pair—A fancy term for a pair of (x, y) coordinates on the normal grid.

x intercept/y intercept—The value where a graph crosses the x axis or y axis.

Defined/undefined—Here's a guaranteed skip for most kids, but as always, if you know the term, the question is very easy. When the ACT asks what values make an expression undefined, they mean when would it violate math rules, either by having a zero on the bottom of a fraction or by having a negative number in a square root. For example, both $\sqrt{-25}$ and $\frac{x-5}{0}$ are undefined. (Don't even show these expressions to your math teacher. She'll get mad. These are the unforgivable curses to math teachers.)

Now, let's take a look at the question from the Pretest:

12. What is the x intercept of the line that contains the points $(-2, 3)$ and $(4, 1)$ in the standard (x, y) coordinate plane?

F. $(\frac{1}{2}, 0)$　　**G.** $(0, \frac{1}{2})$　　**H.** $(7, 0)$　　**J.** $(0, 7)$　　**K.** $(7, \frac{1}{2})$

Solution: In math class, you're usually required to do the algebraic process for a question like this. You could do that here, or you could just sketch a neat diagram. Notice that the answer choices are far enough apart that any half-decent sketch will show you which answer choice is closest and correct. In fact, this question is not really testing your ability to do the algebraic process; it is testing your vocab—x intercept, coordinate plane, and knowing that x comes before y in an ordered pair (x, y). When you know these terms, it's easy.

Correct answer: H

ACT Math Mantra #12
Don't be intimidated by fancy graphing terms.

Fancy Graphing Vocab Drills

Easy

1 Point *P* is to be graphed in a quadrant, not on an axis, of the standard (*x*, *y*) coordinate plane below.

Quadrants of the standard (*x*, *y*) coordinate plane

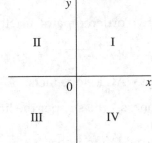

If the *x* coordinate and the *y* coordinate of point *P* are to have the same signs, then point *P* *must* be located in

(A) Quadrant I only

(B) Quadrant III only

(C) Quadrant I or II only

(D) Quadrant II or III only

(E) Quadrant I or III only

Medium

2 In a standard (*x*, *y*) coordinate plane, if one circle is tangent to the *x* axis, and a different circle is tangent to the *y* axis, at how many ordered pairs (*x*, *y*) might the two circles intersect?

(F) 0 only

(G) 1 only

(H) 2 only

(J) 0, 1, or 2

(K) Cannot be determined from the given information

Hard

3 At what value(s) of *x* is $\frac{(x-4)^3}{x^2}$ undefined?

(A) 0 only

(B) 0 and 4 only

(C) −4 only

(D) −4 and 0 only

(E) −4, 0, and 4 only

Green Circle, Black Diamond: Slaloming Slope I

Slope questions appear on every ACT. This topic could fill a college-level course, but the ACT tests only a few concepts.

Most importantly, to find the slope of two ordered pairs, use the *slope formula:*

$$\text{Slope} = \frac{y_1 - y_2}{x_1 - x_2}$$

You should also know these other groovy ACT slope facts:

- The slope of a line measures its steepness—the steeper the line, the bigger the slope.
- Slope is also called *rate* or *rate of change.*
- A line has a positive slope if it rises from left to right.
- A line has a negative slope if it falls from left to right.
- A horizontal line has a slope of 0.

Let's apply this on the question from the Pretest:

13. If the slope of a line through the points $(-2, 5)$ and $(3, b)$ is -1, what is the value of b ?

    A. -2 **B.** -1 **C.** 0 **D.** 1 **E.** 2

Solution: Plug the two points $(-2, 5)$ and $(3, b)$ into the slope equation.

$$\frac{y_1 - y_2}{x_1 - x_2} \Rightarrow \frac{5 - b}{-2 - 3} \Rightarrow \frac{5 - b}{-5} = -1$$

Once you have $\frac{5 - b}{-5} = -1$, you can simply "Use the Answers" and try each answer

choice to see which one makes the equation work. Skill preview: As we will review in Skill 21, you could also cross-multiply to solve this equation.

Correct answer: C

ACT Math Mantra #13
To find the slope or "rate of change" of a line, use $\text{Slope} = \dfrac{y_1 - y_2}{x_1 - x_2}$.

Slaloming Slope I Drills

Easy

1 What is the slope of the line containing the points $(-5, 6)$ and $(-4, -4)$?

Ⓐ 2

Ⓑ $\frac{1}{5}$

Ⓒ 0

Ⓓ $-\frac{1}{5}$

Ⓔ -10

2 What is the slope of the line through the point $(2, -6)$ and the origin?

Ⓕ -3

Ⓖ -2

Ⓗ -1

Ⓙ 0

Ⓚ 4

Medium

3 If the slope of a line through the points $(2, m)$ and $(-2, -5)$ is $\frac{1}{4}$, what is the value of m?

Ⓐ 2

Ⓑ 1

Ⓒ 0

Ⓓ -4

Ⓔ -10

Hard

4 The line graphed below shows the predicted ACT score for a student who masters x number of Skills. Which of the following is the closest estimate of a student's average *rate* of score increase per Skills she or he masters?

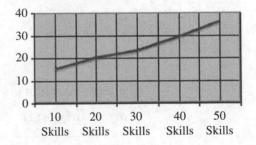

Ⓕ -2

Ⓖ -1

Ⓗ 0

Ⓙ $\frac{1}{2}$

Ⓚ 2

Green Circle, Black Diamond: Slaloming Slope II

Here are the rest of the slope rules on the ACT. You should memorize these; they are on **every** ACT!

• Parallel lines have equal slopes.
• Perpendicular lines have negative reciprocal slopes.

  Example: $\frac{2}{3}$ and $-\frac{3}{2}$

• Skill 32 preview: A line expressed in the form $y = mx + b$ has a slope of m. And $y = mx + b$ is referred to as *slope-intercept form*. If a line is given in standard form $Ax + By = C$, use algebra to convert it to $y = mx + b$ form.

Let's use this on the question from the Pretest:

14. In the standard (x, y) coordinate plane, what is the slope of the line with equation $3x - 5y = -12$?

  F. $\frac{12}{5}$ **G.** 2   **H.** $\frac{3}{5}$ **J.** $-\frac{3}{5}$ **K.** −2

Solution: To find the slope, use algebra to convert $3x - 5y = -12$ to $y = mx + b$ form; i.e., solve the equation for y:

$3x - 5y = -12$ Subtract $3x$ from both sides.
$-5y = -3x - 12$ Divide both sides by -5
$y = \frac{3}{5}x + \frac{12}{5}$

Once the equation is solved for y, the slope is the number in front of the x, in this case $\frac{3}{5}$.

Correct answer: H

ACT Math Mantra #14

An equation of a line in the form $y = mx + b$ is called slope-intercept form, where m represents the slope of the line.

Parallel lines have equal slopes, like $\frac{2}{3}$ and $\frac{2}{3}$.

Perpendicular lines have negative reciprocal slopes, like $\frac{2}{3}$ and $-\frac{3}{2}$.

Green Circle, Black Diamond: Slaloming Slope II Drills

Easy

1 In the standard (x, y) coordinate plane, what is the slope of the line with equation $3y - 9x = 5$?

- Ⓐ -2
- Ⓑ -1
- Ⓒ 0
- Ⓓ 2
- Ⓔ 3

2 In the standard (x, y) coordinate plane, what is the slope of the line with equation $2x + 7y = -14$?

- Ⓕ -2
- Ⓖ $-\dfrac{2}{7}$
- Ⓗ $\dfrac{2}{7}$
- Ⓙ 2
- Ⓚ 7

Medium

3 What is the slope-intercept form of $6x - 2y - 4 = 0$?

- Ⓐ $y = -3x + 4$
- Ⓑ $y = 3x + 4$
- Ⓒ $y = 3x - 2$
- Ⓓ $y = -3x - 2$
- Ⓔ $y = -x + 4$

4 What is the slope of the line that is perpendicular to the line through the points $(0, -3)$ and $(-2, -2)$?

- Ⓕ -1
- Ⓖ 0
- Ⓗ 1
- Ⓙ 2
- Ⓚ Undefined

Hard

5 If the slope of a line through the origin and $(-2, b)$ is 2, and the line through $(b, -2)$ and $(a, 5)$ is parallel to that line, what is the value of a ?

- Ⓐ 1
- Ⓑ 0.5
- Ⓒ 0
- Ⓓ -0.5
- Ⓔ -1

The Sports Page: Using Charts and Graphs

Charts and graphs display information, just like the sports page lists a team's wins and losses. In fact, using charts and graphs should feel no different than checking stats for your favorite Red Sox pitcher, Halo III high-scorer, or Guitar Hero hero.

ACT Math Mantra #15
The key to charts and graphs is to read the intro material and the "note"
if there is one, and to expect an average, percent, and/or probability
question about the data.

The ACT uses several kinds of charts and graphs. Charts display information in rows and columns. Pie graphs represent information as part of a pie. Line graphs display how data changes, often over time. Bar graphs compare the values of several items, such as sales of different toothpastes. Pictographs use small pictures to represent data. The key to pictographs is noticing the legend. If each icon represents 8 books, then $^1/_2$ an icon represents 4 books, not $^1/_2$ of a book.

The key to any of these is to stay relaxed and just read the data that the chart or graph displays. Don't get intimidated. If the chart is unusual, the ACT will explain how to read it in an intro or "note." Also, we can predict most questions. Almost always the questions ask you to find the average of the data in the table, the percent that one of the categories occupies of the whole, and/or the probability of a certain piece of data occurring.

Let's take a look at the question from the Pretest:

The following chart shows the results when 31 high school juniors were asked to write the name of their favorite movie on a slip of paper and place it in a box.

Movie	Votes
Wedding Crashers	4
Across the Universe	6
Once	2
Lord of the Rings trilogy	5
Harry Potter movies	6
Superbad	3
Other	5

15. If a slip of paper is chosen at random from the box, which of the following is closest to the percent chance that the slip chosen will name *Wedding Crashers*?

   A. 4%   B. 13%   C. 31%
   D. 35%   E. 69%L

Solution: Easy. The percent chance of choosing a slip with *Wedding Crashers* written on it is the number of slips that say that, divided by the total slips: $\frac{4}{31}$, which equals 0.129 or approximately 13%.

Correct answer: B

The stem-and-leaf plot below shows the number of times Willy grabbed a rebound in each of 14 basketball games.

Stem	Leaf
0	7 9
1	1 2 2 2 3 5 7 9 9
2	1 2 3

(*Note:* For example, 12 rebounds would have a stem value of 1 and a leaf value of 2.)

Medium

❶ Which of the following is closest to the mean number of rebounds Willy grabbed per game?

Ⓐ 13

Ⓑ 14

Ⓒ 15

Ⓓ 16

Ⓔ 17

New Leaf Learning Center held its annual yogathon for 3 days. The total money raised in the 3 days was $24,500. The money raised, in dollars, is shown for each of the 3 days in the bar graph below.

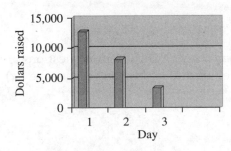

❷ Approximately what percent of the money raised by the yogathon over the 3 days did New Leaf Learning Center raise on day 2 ?

Ⓕ 12%

Ⓖ 25%

Ⓗ 33%

Ⓙ 50%

Ⓚ 65%

The Only Function Questions on the ACT

Most students fear functions. "I suck at functions," I hear from almost every new student. I'm not sure where this attitude comes from, but here's the good news: functions on the ACT are **easy**. While functions could be the topic of a full-year university course, the ACT only tests **one function concept**. That's right, only one!

First, let's get rid of function phobia. Functions are just a type of equation, like $y = mx + b$. To show that an equation is a function, sometimes people replace the "y" with "$f(x)$" or "$g(x)$" or "$h(x)$." That's it. To solve functions, remember that "$f(x)$" is just a fancy way of saying "y." So, $f(x) = 2x - 1$ means the same as $y = 2x - 1$.

The one function concept that the ACT tests is: "plug whatever is in the parentheses in for x."

ACT Math Mantra #16

$f(3)$ means "plug 3 in for x."

$f(m)$ means "plug m in for x."

$f(g(m))$ means "plug $g(m)$ in for x."

That's it. Whatever is in the parentheses, you plug in for x! No thinking. No tricks. No curveballs. On the ACT, just plug whatever is inside the parentheses into the equation for x. That's all for ACT functions.

Now, let's use this on the question from the Pretest:

16. If $f(x) = -2x^2 - 3$, what is $f(-3)$?

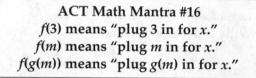

    F. 33   **G.** 15   **H.** 3   **J.** −21   **K.** −39

Solution: This question simply asks you to plug the −3 in for x in the equation. Cake!

So

$f(-3) = -2(-3)^2 - 3$ Plug −3 in for x.
        $-2(9) - 3$
        $-18 - 3$
        -21

Correct answer: J

Repeat after me, "I do not fear functions. I enjoy them. ACT function questions are fun, easy, and interesting." As long as you don't get intimidated when you see the $f(x)$, ACT functions are easy!

The Only Function Questions on the ACT Drills

Easy

① If $f(x) = 2x^2 + 12$, what is $f(3)$?

(A) 3

(B) 12

(C) 18

(D) 30

(E) 48

Medium

② Let a function of 2 variables be defined by $f(x, y) = 2x - y(4 - x)$. What is the value of $f(4, 8)$?

(F) −4

(G) 0

(H) 3

(J) 5

(K) 8

③ The number of friends that Matty had on Facebook for the first 60 days after he joined can be modeled by the function $F(d) = 14d + 50$, where $d = 0$ corresponds to the day he joined. Using this model, how many friends would you expect him to have 39 days after joining?

(A) 546

(B) 596

(C) 840

(D) 890

(E) 904

Hard

④ Given $f(x) = 2x^2 + 3x - 4$, what is $f(x + k)$?

(F) $2x^2 + k^2 + 3x + 3k - 4$

(G) $2x^2 + 2k^2 + 3x + 3k - 4$

(H) $2(x + k)^2 + 3xk - 4$

(J) $2(x + k)^2 + 3x + 3k - 4$

(K) $2x^2 + 3x - 4 + k$

⑤ A function is defined by $h(x) = -x^4$. Which of the following is an expression for $h(h(x))$?

(A) Undefined

(B) x^{16}

(C) x^4

(D) $-x^4$

(E) $-x^{16}$

Midpoint and Distance Formulas

You couldn't ask for easier points to boost your score. Every ACT has a midpoint and/or distance question. If you don't know the formulas, you don't stand much chance. But if you memorize them, right here and now, you will gain points, guaranteed! In fact, take a few minutes right now to cut out the flashcards from the back of this book. Bring them everywhere you go, school, sports, DMX concerts, parties, etc. Everyone loves math flash cards!

To find the midpoint between two points on a graph, use the midpoint formula. All this formula really says is, "Take halfway between the x numbers and halfway between the y numbers." Halfway between two numbers is the same as the average of the numbers, so the midpoint formula is like the average formula:

$$\text{Midpoint} = \left(\frac{x + x}{2}, \frac{y + y}{2} \right)$$

The distance formula also makes sense if you really look at it. It is based on the Pythagorean theorem (Skill 24). To find the distance between two points on a graph, use the formula

$$\text{Distance} = \sqrt{(x - x)^2 + (y - y)^2}$$

Cut out the flashcards at the end of this book to help you memorize these formulas.

ACT Math Mantra #17
To find the midpoint of two points, use the formula: Midpoint $= \left(\frac{x + x}{2}, \frac{y + y}{2} \right).$

To find the distance between two points use the formula:

$$\textbf{Distance} = \sqrt{(x - x)^2 + (y - y)^2}.$$

Let's use one of these on the question from the Pretest:

17. A line in the standard (x, y) coordinate plane contains the points $P(3, -5)$ and $Q(-7, 9)$. What point is the midpoint of $\overline{PQ}$?

    A. $(-2, 4)$ **B.** $(-2, -4)$ **C.** $(-2, 2)$ **D.** $(-2, -2)$ **E.** $(3, 3)$

Solution: A midpoint is halfway between two points, really just the average. That's what the midpoint formula gives us, the average of the two points:

$$\left(\frac{3 + (-7)}{2}, \frac{-5 + 9}{2} \right) = (-2, 2)$$

Correct answer: C

Midpoint and Distance Formulas Drills

Easy

1 The endpoints of $\overline{PQ}$ on the real number line are −14 and 2. What is the coordinate of the midpoint of $\overline{PQ}$?

(A) −8

(B) −6

(C) −2

(D) 0

(E) 8

2 A line in the standard (x, y) coordinate plane contains the points $M(-2, 4)$ and $N(8, 10)$. What point is the midpoint of $\overline{MN}$?

(F) (−2, 4)

(G) (3, 3)

(H) (5, 3)

(J) (5, 7)

(K) (3, 7)

Medium

3 In the standard (x, y) coordinate plane, what is the distance between the points (−4, 2) and (1, −10) ?

(A) 3

(B) 5

(C) 9

(D) 11

(E) 13

4 A diameter of a circle has endpoints (−2, 10) and (6, −4) in the standard (x, y) coordinate plane. What point is the center of the circle?

(F) (2, 3)

(G) (4, 6)

(H) (6, 4)

(J) (12, 8)

(K) Cannot be determined from the given information

Hard

5 A city's restaurants are laid out on a map in the standard (x, y) coordinate plane. How long, in units, is the straight-line path between Paul & Elizabeth's restaurant at (17, 15) and The Green Bean restaurant at (12, 5) ?

(A) $\sqrt{429}$

(B) $\sqrt{324}$

(C) $\sqrt{125}$

(D) 11

(E) $\sqrt{15}$

Make It Real

Dr. Evil: Finally we come to my number two man. His name? Number Two.
Austin Powers: International Man of Mystery **(New Line Cinema, 1997)**

If "Use the Answers" is our single most useful and important strategy, "Make It Real" is number two. You could say that if "Use the Answers" is Dr. Evil, then "Make It Real" is . . . well, Number Two.

Let's take a look at the question from the Pretest:

18. If $c = -2b - 2(2e - b)$, what happens to the value of c when b and e are both increased by 2 ?

    F. It is unchanged.   **G.** It is increased by 2.   **H.** It is increased by 4.
    J. It is decreased by 4.   **K.** It is decreased by 8.

Solution: You'd have to be Einstein to get this question; it's too theoretical. So we take it out of theory and "Make It Real." We choose real numbers in place of the variables, and it becomes easier. We can choose any numbers that seem to fit the equation, and we might as well choose numbers that work evenly and avoid decimals. In this case, let's say $b = 3$ and $e = 4$. (Be careful of choosing 0, 1, or 2 for "Make It Real." These numbers behave in strange ways, as we'll see in Skill 46.) So when $b = 3$ and $e = 4$:

$$c = -2b - 2(2e - b)$$
$$c = -2(3) - 2(2(4) - 3)$$
$$c = -6 - 2(5)$$
$$c = -6 - 10$$
$$c = -16$$

The question asks what happens when we add 2 to both b and e, so let's add 2 to both b and e and then solve the equation again. With $b = 5$ and $e = 6$:

$$c = -2b - 2(2e - b)$$
$$c = -2(5) - 2(2(6) - 5)$$
$$c = -10 - 2(7)$$
$$c = -10 - 14$$
$$c = -24$$

Cake. C was −16 and is now −24, so it decreased by 8.

Correct answer: K

This strategy turns very difficult questions into much easier ones! It's my single favorite ACT strategy.

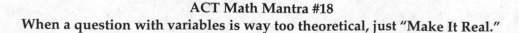

ACT Math Mantra #18
When a question with variables is way too theoretical, just "Make It Real."

Make It Real Drills

Easy

1 If *t* represents an odd integer, which of the following expressions also represents an odd integer?

(A) $t - 3$

(B) $2t - 1$

(C) $2t - 2$

(D) $3t + 1$

(E) $4t + 4$

Medium

2 If $m < 0$ and $n > 0$, then the product of *m* and *n*

(F) Is always even

(G) Is always odd

(H) Is always positive

(J) Is always negative

(K) Cannot be zero, but can be any other real number

3 If *b* is any real number other than 3 and 7, then $\dfrac{(3 - b)(b - 7)}{(b - 3)(b - 7)} =$

(A) 21

(B) 1

(C) 0

(D) −1

(E) −21

Hard

4 There are *k* students in a class. If, among those students, *q*% have NOT seen the movie *Monty Python and the Holy Grail*, which of the following general expressions represents the number of students who have seen the movie?

(F) kq

(G) $0.01kq$

(H) $\dfrac{(100 - q)k}{100}$

(J) $\dfrac{(1 - q)k}{0.01}$

(K) $100(1 - q)k$

5 Let *x* equal $-3m - 2n - 4$. What happens to the value of *x* if the value of *m* increases by 2 and the value of *n* decreases by 1 ?

(A) It increases by 3.

(B) It increases by 2.

(C) It is unchanged.

(D) It decreases by 3.

(E) It decreases by 4.

Perimeter, Area, Volume

For the ACT you need to have certain formulas memorized. You need to know these:

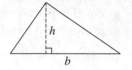

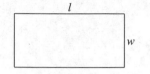

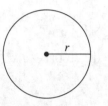

Area of a triangle = $\frac{1}{2}$(base)(height)

Area of a rectangle = (length)(width)
= (base)(height)

Area of a parallelogram = (base)(height)

Area of a circle = πr^2
Circumference of a circle = $2\pi r = \pi d$

Volume of a solid = (length)(width)(height)

And the perimeter of any shape is the addition of the lengths of the sides. When they want you to use a more uncommon perimeter, area, or volume formula, they'll give it to you in the question.

ACT Math Mantra #19
Memorize basic perimeter, area, and volume formulas.
To use them, plug in what you know, and solve for the variable.

Let's take a look at the question from the Pretest:

19. A square lot, with area 625 square feet, is completely fenced. What is the length, in feet, of the fence?

   A. 25
   B. 75
   C. 100
   D. 125
   E. 300

Solution: Since the square has area 625, we can determine the length of a side of the square. Set up the area formula. Plug in what we know and solve for the variable. This is a great ACT strategy! So $(s)(s) = s^2 = 625$. Take the square root of both sides to get $s = 25$. Since one side measures 25, the perimeter is $(4)(25) = 100$. Make sure to finish the question. Don't be tempted by choice A. Ask yourself, "Self, did I finish the question?" With practice you can avoid this number one most common ACT careless error.

Correct answer: C

Perimeter, Area, Volume Drills

Easy

① A square lot, with 75-foot sides, is completely fenced. What is the approximate length, in feet, of the fence?

- Ⓐ 70
- Ⓑ 75
- Ⓒ 150
- Ⓓ 225
- Ⓔ 300

Medium

② If 10,000 cubic yards of water is to be funneled into a rectangular tank with a floor that measures 75 yards by 110 yards, about how many yards deep will the water be?

- Ⓕ More than 4
- Ⓖ Between 3 and 4
- Ⓗ Between 2 and 3
- Ⓙ Between 1 and 2
- Ⓚ Less than 1

③ Parallelogram *MNOP*, with dimensions in centimeters, is shown in the diagram below. What is the area of the parallelogram, in square centimeters?

- Ⓐ 300
- Ⓑ 480
- Ⓒ 544
- Ⓓ 600
- Ⓔ 680

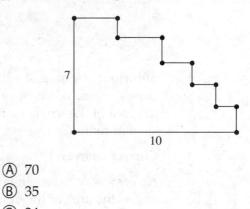

④ A square is circumscribed about a circle of 9-inch radius, as shown below. What is the area of the square in square inches?

- Ⓕ 81
- Ⓖ 144
- Ⓗ 81π
- Ⓙ 308
- Ⓚ 324

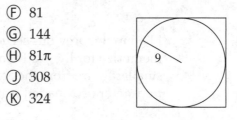

⑤ In the figure below, each pair of intersecting line segments meets at a right angle. What is the perimeter of the figure?

- Ⓐ 70
- Ⓑ 35
- Ⓒ 34
- Ⓓ 17
- Ⓔ Cannot be determined from the given information

Donuts

"Mmmm . . . donuts."
Homer Simpson, *The Simpsons* (Fox, 1992)

I lied in the previous section. There is one more important formula that you need to memorize for the ACT: To find the **area** of a shaded region, just subtract the area of the smaller figure from the area of the larger figure. You might also be asked to find the **perimeter** of a shaded region.

Let's look at the question from the Pretest:

20. In the diagram below, if the radius of the large circle is 6 and the radius of the small circle is 4, what is the area of the shaded region?

F. 2
G. 20
H. 2π
J. 20π
K. $\sqrt{20}$ M

Solution: The area of a shaded region is found by subtracting the area of the little guy from the area of the big guy. The area of the large circle = $\pi(r^2) = \pi(6^2) = \pi(36)$. And the area of the smaller circle = $\pi(r^2) = \pi(4^2) = \pi(16)$. So $36\pi - 20\pi = 16\pi$ is the area of the shaded region.

Correct answer: J

An easy way to remember this is, "The area of a donut equals the area of the big guy minus the area of the donut hole." This is a great formula. Almost every ACT has one question using it.

ACT Math Mantra #20
The area of a donut equals the area of the big guy
minus the area of the donut hole.

Donuts Drills

Easy

❶ Openings for 3 square windows, each 2 feet on a side, were cut from a rectangular wall 7 feet by 10 feet. What is the area, in square feet, of the remaining portion of the wall?

Ⓐ 70

Ⓑ 66

Ⓒ 62

Ⓓ 58

Ⓔ 54

Medium

❷ A square is circumscribed about a circle of 9-inch radius, as shown below. What is the area, in square inches, of the region that is inside the square but outside the circle?

Ⓕ $81 - 9\pi$

Ⓖ $144 - 9\pi$

Ⓗ $324 - 18\pi$

Ⓙ $308 - 81\pi$

Ⓚ $324 - 81\pi$

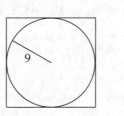

❸ In the figure below, the area of rectangle $ABCD$ is 48 square units. What is the area of the shaded trapezoid, in square units?

Ⓐ 48

Ⓑ 35

Ⓒ 30

Ⓓ 24

Ⓔ 12

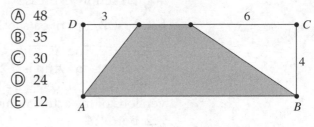

Hard

❹ The larger square in the figure below has 8-inch sides and circumscribes the smaller square. If the larger square intersects the smaller square at the four midpoints of the sides of the larger square, what is the area, in square inches, of the shaded region?

Ⓕ 16

Ⓖ 24

Ⓗ 32

Ⓙ 48

Ⓚ 64

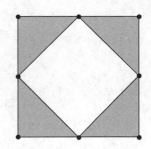

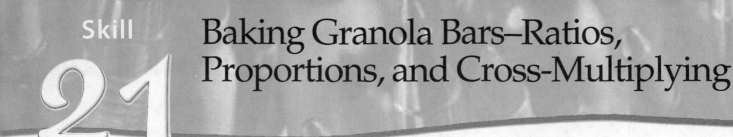
Baking Granola Bars–Ratios, Proportions, and Cross-Multiplying

A ratio simply expresses a relationship between two numbers, such as to bake granola bars, use 7 cups of oats to 2 cups of maple syrup. This can be written as 7:2 or $\frac{7}{2}$ or even 7 to 2. The ACT likes to see if you can play with the ratios. For example, since the ratio of oats to syrup is 7:2, we could also say that the ratio of oats to the oats/syrup combo is 7:9, since $7 + 2 = 9$. Also, remember that the numbers in a ratio may be reduced versions of the actual numbers. For example, if we bake a double batch, we will use not 7:2, but 14:4.

Two ratios that are equal, such as $\frac{5}{12} = \frac{10}{24}$, are called a *proportion*. On the ACT, if one of the four numbers in that proportion is unknown, use cross-multiplying to solve for it. For example, if $\frac{5}{12} = \frac{x}{40}$ and you want to find the value of x, just cross-multiply: $(5)(40) = (12)(x)$. Then use algebra to solve for x (divide both sides by 12 to get $x = 16.66$). **Anytime the ACT shows you a proportion, cross-multiply.** It's totally predictable. I love things we can absolutely predict on the ACT! So when you come to a proportion on the ACT, get excited, stand up from your desk, and say, "I love the ACT!" People might look at you, but with our mantra, you will **always** get the question right!

ACT Math Mantra #21
4 boys to 5 girls could also be expressed as 5 girls to 9 students.
A ratio can be a reduced version of the real numbers.
And when you see a proportion on the ACT, cross-multiply.

Let's look at the question from the Pretest:

21. A triangle with a perimeter of 89 has one side that is 19 inches long. The lengths of the other two sides have a ratio of 3:7. What is the length, in inches, of the *longest* side of the triangle?

   A. 19   B. 21   C. 29   D. 40   E. 49

Solution: If one side of the triangle is 19 and the perimeter is 89, then the other two sides add up to $89 - 19 = 70$. One of these must be the *largest* side, and we know that these two sides must have a ratio of $3:7$. Rewrite the ratio as part : total instead of part : part, and set up the proportion:

$$\frac{\text{larger side}}{\text{sum of two sides}} \Rightarrow \frac{x}{70} = \frac{7}{10}.$$

Then cross-multiply to get $10x = 490$. Divide both sides by 10 and $x = 49$.

Correct answer: E

Baking Granola Bars Drills

Easy

1 On the line segment below, the ratio of lengths *MN* to *MO* is 2:3. What is the ratio of *MN* to *NO* ?

(A) 1:2

(B) 2:1

(C) 1:3

(D) 3:1

(E) Cannot be determined from the given information

M • ———————— N • ———— O •

2 At a glassworks, 60 pounds of composite is required to produce 15 pounds of glass. How many pounds of composite are required to produce 3 pounds of this glass?

(F) 12

(G) 18

(H) 24

(J) 36

(K) 45

Medium

3 On the blueprint for Shantel's house, $\frac{1}{8}$ inch represents an actual length of 1 foot. What is the area, in square feet, of Shantel's living room, which is a rectangle $2\frac{1}{8}$ inches by 2 inches on the blueprint?

(A) 33

(B) 34

(C) 68

(D) 136

(E) 272

Hard

4 The ratio of *x* to *y* is 2 to 5, and the ratio of *z* to *y* is 3 to 4. What is the ratio of *x* to *z* ?

(F) 2 to 3

(G) 5 to 9

(H) 7 to 8

(J) 8 to 15

(K) 9 to 2

Intimidation and Easy Matrices

Many kids think matrices are super high-level math that they could never get, but the ACT asks only the most basic matrices questions. First, let's get the stress out. Have a good scream, hug it out, and we'll move on.

Mostly, the ACT uses matrices like ordinary charts, like the charts in Skill 15. Reading these is no different than checking the sports page. In fact, if these questions just omitted the word *matrix*, more kids would try them and get them right. Just including the word *matrix* in a question can change its ranking from "easy" to "hard."

There are a few things that the ACT might actually ask you to do with matrices.

- The first is to add matrices. This is so easy that you barely do it in school. To add matrices, you just add the numbers in each location. You'll see this in the drills.
- The second thing they could ask is for you to multiply matrices. They ask this very rarely, but the key when they do is to know that when you multiply two matrices, the result will have as many rows as the first and as many columns as the second matrix. For example, a **2-row** by 3-column matrix times a 3-row by **4-column** matrix will have **2 rows and 4 columns**. And the middle numbers (3s in this case) must match or the matrices can't be multiplied.
- That's about it for big ol' scary matrices. If the ACT asks you to do anything else, they will tell you how to do it in the question. They won't expect you to have anything else memorized.

Let's use this on the question from the Pretest:

22. The number of students in three school clubs is shown in the following matrix.

$$\begin{bmatrix} \text{Yoga} & \text{Spanish} & \text{Best Buddies} \\ 20 & 18 & 36 \end{bmatrix}$$

The school paper reported that the student body of the school was comprised according to the ratios below. Given these matrices, if participation in clubs is spread evenly among the grades, how many freshmen could be estimated to be in the Yoga Club?

F. 0.2
G. 0.3
H. 20
J. 10
K. 6

$$\begin{bmatrix} 0.3 - \text{Freshmen} \\ 0.5 - \text{Sophomores \& Juniors} \\ 0.2 - \text{Seniors} \end{bmatrix}$$

Solution: This question definitely scares most kids away. But if we just drop the words *matrix* and *matrices*, it's a normal chart question. The first chart shows how many kids are in each club, and the second chart shows how the student body is divided by grades, i.e., the freshmen are 0.3 or "three-tenths" of the school. So 0.3 of the total of 20 kids in the Yoga Club must be frosh: $(0.3)(20) = 6$ freshmen in the Yoga Club.

Easy, right? Some kids think matrices belong in college-level Calc III, but on the ACT they are no problem!

Correct answer: K

ACT Math Mantra #22
Relax when you see a matrix question; just treat the matrix like a normal chart. To add matrices, add corresponding numbers; to multiply, remember that the result will have as many rows as the first and as many columns as the second matrix being multiplied; and for any other operation, just follow the instructions that the question provides.

Intimidation and Easy Matrices Drills

Medium

❶ By definition the determinant $\begin{vmatrix} a & b \\ c & d \end{vmatrix}$ equals $ad - bc$. What is the value of $\begin{vmatrix} 3m & 2m \\ 4n & 3n \end{vmatrix}$ when $m = -2$ and $n = 3$?

Ⓐ −6

Ⓑ −13

Ⓒ −55

Ⓓ −96

Ⓔ −102

❷ In a recent high school election, the number of votes for three candidates is shown in the following matrix.

$$\begin{bmatrix} \text{Angino} & \text{Haas} & \text{Bloomberg} \\ 50 & 48 & 48 \end{bmatrix}$$

The school paper reported that the winner received votes in the ratios shown below.

Given these matrices, what is the number of juniors who voted for the winner?

Ⓕ 0.2

Ⓖ 0.3

Ⓗ 25

Ⓙ 15

Ⓚ 10

$$\begin{bmatrix} 0.3\text{—Freshmen} \\ 0.5\text{—Sophomores} \\ 0.2\text{—Juniors} \end{bmatrix}$$

❸ The matrix below shows the lights at a stadium (sections A through H) that are on and off, with 1 representing on and 0 representing off. Based on this information and the information in the matrix, which of the following is true?

Ⓐ All lights in sections A to H are on.

Ⓑ All lights in sections A to H are off.

Ⓒ The lights in sections A, C, and H are off.

Ⓓ The lights in sections A, C, and H are on.

Ⓔ There are five lampposts at the stadium.

$$\begin{bmatrix} \text{A} & \text{B} & \text{C} & \text{D} & \text{E} & \text{F} & \text{G} & \text{H} \\ 0 & 1 & 0 & 1 & 1 & 1 & 1 & 0 \end{bmatrix}$$

❹ $\begin{bmatrix} p & q \\ r & s \end{bmatrix} + \begin{bmatrix} w & x \\ y & z \end{bmatrix} = ?$

Ⓕ $\begin{bmatrix} pw & qx \\ ry & sz \end{bmatrix}$

Ⓖ $\begin{bmatrix} 11 \\ 11 \end{bmatrix}$

Ⓗ $\begin{bmatrix} p+w & q+x \\ r+y & s+z \end{bmatrix}$

Ⓙ $\begin{bmatrix} p-w & q-x \\ r-y & s-z \end{bmatrix}$

Ⓚ Undefined

Art Class

When you are given a diagram on the ACT, ask yourself if it seems accurate. If it does, you can use it, sometimes just to see what to do next, and other times to get a correct answer without even doing much math. For example, if you are given the length of some part of the diagram, you can often use that to estimate an unknown. You've done this before. Imagine you're on a road trip. You look at the map and say, "We have to go from here to here on this squiggly highway." The map key says that each inch is 100 miles, so you use your thumb to represent an inch and you estimate the length of the squiggly line highway. We call this "Use the Diagram."

If the diagram is clearly **not** drawn to scale (not accurate), resketch it somewhat accurately and then "Use the Diagram." It turns out that often the whole question is hinged on its being out of scale, and when you put it into scale, the answer becomes obvious.

Also, while we're in art class, here's another great strategy. If a question describes a diagram, but none is shown, draw one. Sometimes this gives the answer immediately, and sometimes it shows you what to do next, but either way it always helps!

At first all this might feel weird. For a year in geometry class you were taught not to estimate with a diagram. Plus, on the ACT, estimating makes the question so much easier that it feels "cheap," like you are cheating. It's not! It's actually what they want you to do. Remember the test is supposed to test your cleverness, not just what you learned in math class. This strategy brings out your innate cleverness.

Let's look at the Pretest question:

23. Points A, B, C, D, and E are points on a line in that order. If B is the midpoint of $\overline{AC}$, C is the midpoint of $\overline{AD}$, and D is the midpoint of $\overline{AE}$, which of the following is the longest segment?

    A. $\overline{AB}$ **B.** $\overline{AD}$ **C.** $\overline{BC}$ **D.** $\overline{BD}$ **E.** $\overline{CE}$

Solution: Sketch the diagram to scale, following the instructions in the question:

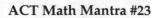

Now the answer is obvious. Of the choices, $\overline{CE}$ is longest.

Correct answer: E

ACT Math Mantra #23
"Use the Diagram" to estimate an answer. When a diagram is not drawn to scale, redraw it. And when a picture is described, but not shown, draw it! When estimating an answer, translate fractions, $\sqrt{\ }$, or π into decimals.

Art Class Drills

Easy

① If points A, B, C, and D are different points on a line; $AB = BC = 8$; and D is between B and C; which of the following could be the measure of segment AD ?

(A) 3
(B) 5
(C) 8
(D) 10
(E) 16

Medium

② What is the y intercept of the line in the standard (x, y) coordinate plane that goes through the points $(-4, 5)$ and $(4, 1)$?

(F) −1
(G) 1
(H) 3
(J) 5
(K) 7

Hard

③ In the figure below, A, B, and C are collinear, and angles A, C, and EBD are right angles. If $EB = 14$, $DB = 7$, and $DC = 6$, to the nearest whole number, what is the measure of EA ?

(A) 3
(B) 5
(C) 7
(D) 10
(E) 16

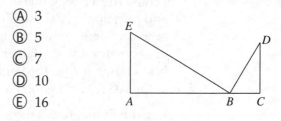

④ A room has the shape and dimensions in meters given below. A support beam is located halfway between point P and point Q. Which of the following is the distance of the beam from point M ?

(F) 3
(G) $\sqrt{15.25}$
(H) $\sqrt{36.75}$
(J) $\sqrt{51.25}$
(K) 9

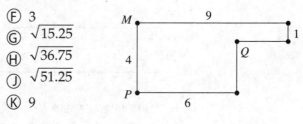

The Mighty Pythagoras

In Skills 5 to 7, we solved for missing angles in a diagram. Skills 24 to 26 show you how to solve for a missing side in a diagram.

I come from a long line of teachers. My grandfather Herman was a beloved teacher and principal in Zshmolphf, Austria. His father was a scholar. His father's father was a scholar. Some say we can trace our lineage back thousands of years to the James Brown of geometry, the godfather of good math, your friend and mine, the mighty Pythagoras, best known for his theorem $a^2 + b^2 = c^2$. Pythagoras is also definitely a favorite of the folks at the ACT. His theorem is used on **Every** ACT.

Use the Pythagorean theorem, $a^2 + b^2 = c^2$, to find the length of a side of a right triangle when you are given the lengths of the other two. The letters a and b always represent the shorter sides (called legs), and c always represents the longest side, which is opposite the right angle. The longest side is also called the hypotenuse. As I said, this is used on **Every** ACT. In fact, anytime you see a right triangle on the ACT, you will probably need it. Just plug the two sides that you know into the equation, and solve for the third. Here's an example:

If $a = 8$ and $c = 17$, what is the value of b?

$$a^2 + b^2 = c^2$$
$$8^2 + b^2 = 17^2$$
$$64 + b^2 = 289$$
$$b^2 = 225$$
$$b = 15$$

ACT Math Mantra #24
When you see a right triangle, try $a^2 + b^2 = c^2$.

Let's use good ol' Pythagoras on the question from the Pretest:

24. In a right triangle, the measures of two sides are 6 and 10, which of the following could be the measure of the third side?

    F. 3   **G.** 4   **H.** 6   **J.** 8   **K.** 10

Solution: Great review of ACT Math Mantra #23, "When a picture is described but not shown, draw it." This helps you visualize and organize the info and shows you what to do next. The question does not state whether 10 is another short side like 6 or if it is the longest side. But we can "Use the Answers." We simply try each answer choice in $a^2 + b^2 = c^2$, always using the biggest number for c. Choice A is not correct since $3^2 + 6^2 \neq 10^2$, and choice B is not correct since $4^2 + 6^2 \neq 10^2$, and so on. Choice D is correct since $6^2 + 8^2 = 10^2$.

Remember that the longest side, the one opposite the right angle, the one our buddy Pythagoras called c, also goes by the nickname *hypotenuse*. So if you see the word *hypotenuse*, just know that it's a fancy vocab word for the side opposite the right angle.

Correct answer: J

The Mighty Pythagoras Drills

Easy

❶ Jack leaves the caves and runs directly north for 6 miles. He then turns and runs 8 miles east. How many miles is he from the caves?

Ⓐ 2
Ⓑ 8
Ⓒ 10
Ⓓ 14
Ⓔ 100

Medium

❷ If a rectangle measures 27 yards by 36 yards, what is the length, in yards, of the diagonal of the rectangle?

Ⓕ 24
Ⓖ 27
Ⓗ 36
Ⓙ 45
Ⓚ 63

❸ Points L, M, and N are shown in the standard (x, y) coordinate plane below. If the three points are connected to form a triangle, what is the measure of the hypotenuse of the triangle in coordinate units?

Ⓐ $2\sqrt{10}$
Ⓑ $4\sqrt{2}$
Ⓒ 4
Ⓓ 3
Ⓔ 2

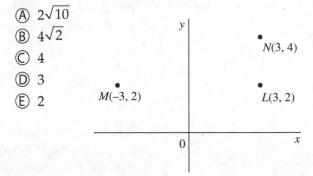

Hard

❹ In the figure below, $\triangle ABC$ is a right triangle with legs that measure m and $3m$ feet, respectively. What is the length, in feet, of the hypotenuse?

Ⓕ $\sqrt{10}m$
Ⓖ $\sqrt{7}\,m$
Ⓗ $\sqrt{3}\,m$
Ⓙ $2m$
Ⓚ $4m$

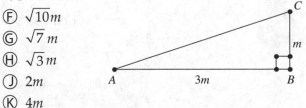

Special Right Triangles

When the Pythagorean theorem is not enough to find the length of a missing side, use one or both of the two "special right triangles:"

❶ When the three angles of a triangle measure 30°, 60°, 90°, then the sides are x, $x\sqrt{3}$, and $2x$.

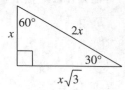

❷ When the three angles measure 45°, 45°, 90° (also called an isosceles right triangle), then the two short sides are equal, let's call them x, and the longest side measures $x\sqrt{2}$. Or, if you are given the long side, then the two short sides

each measure $\frac{x}{\sqrt{2}}$.

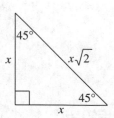

ACT Math Mantra #25
When you see a 30°, 45°, or 60° angle in a right triangle, try using the special right triangles.

Now, let's use these on the question from the Pretest:

25. What is the shortest side of a triangle that is congruent to triangle *PQR* shown below?

A. 3
B. $3\sqrt{3}$
C. 6
D. $6\sqrt{3}$
E. 12

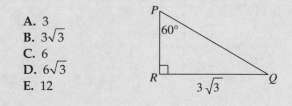

Solution: When you see a right triangle on the ACT, first try the Pythagorean theorem. Given only one side of the triangle, we don't have much info to use in the Pythagorean theorem, so try the special right triangles. **Also the 60° angle is a clue to use special right triangles. When you see a 60° angle, try special right triangles.** Since there is a 90° and a 60° angle, the third angle must be 30°, so it is a 30, 60, 90 triangle and follows the pattern x, $x\sqrt{3}$, and $2x$ for the three sides. Therefore, *PR*, the shortest side, equals 3. Since the two triangles are congruent, the measure of the shortest side of the other triangle must also be 3.

Correct answer: A

Special Right Triangles Drills

Medium

1 Which of the following sets of 3 numbers could be the side lengths, in meters, of a 30°, 60°, 90° triangle?

Ⓐ $3, 3, 3$

Ⓑ $3, 3, 3\sqrt{2}$

Ⓒ $3, 3\sqrt{2}, 3\sqrt{2}$

Ⓓ $3, 3\sqrt{2}, 3\sqrt{3}$

Ⓔ $3, 3\sqrt{3}, 6$

2 In the figure below, $WXYZ$ is a square and M, N, O, and P are the midpoints of its sides. If $WX = 8$ centimeters, what is the perimeter of $MNOP$, in centimeters?

Ⓕ 72

Ⓖ $36\sqrt{2}$

Ⓗ $24\sqrt{2}$

Ⓙ $16\sqrt{2}$

Ⓚ 16

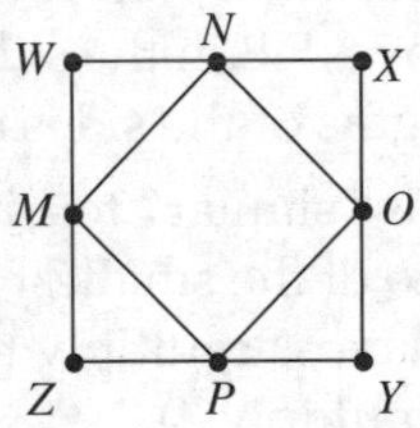

3 What is the length of the longest leg of a triangle congruent to the triangle shown below?

Ⓐ 3

Ⓑ $3\sqrt{3}$

Ⓒ 6

Ⓓ $6\sqrt{3}$

Ⓔ 12

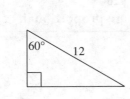

Hard

4 In an isosceles right triangle, the sum of measures of the two equal sides is 9. What is the measure of the longest side?

Ⓕ 4.5

Ⓖ $4.5\sqrt{2}$

Ⓗ 9

Ⓙ $4.5\sqrt{3}$

Ⓚ $9\sqrt{3}$

5 In a 30, 60, 90 triangle, the sum of measures of the shortest and longest sides is 12. What is the measure of the third side?

Ⓐ 12

Ⓑ $12\sqrt{3}$

Ⓒ $8\sqrt{3}$

Ⓓ $4\sqrt{3}$

Ⓔ $\sqrt{3}$

Dr. Evil and Mini-Me

I shall call him . . . "Mini-Me."

Dr. Evil, *Austin Powers: The Spy Who Shagged Me* (New Line Cinema, 1999)

Similar triangles are two triangles where one is a shrinky version of the other, like Dr. Evil and Mini-Me. Since one is a shrunken version of the other, all sides are proportional, shortest to shortest and longest to longest. The ACT loves similar triangles.

Let's have a look at the question from the Pretest:

26. If triangle *MNO* (not shown) is similar to triangle *PQR* shown below, and has a shortest side with length 4.5, which of the following would be the measure of longest side of triangle *MNO*?

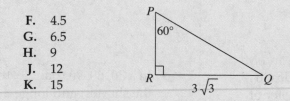

F.   4.5
G.   6.5
H.   9
J.   12
K.   15

Solution: Here's your chance to see if you really learned Skill 25. We have to determine the measures of sides of triangle *PQR* shown above. Fill in the sides of special right triangle $\triangle PQR$: 3, $3\sqrt{3}$, 6. Once you have the lengths of the sides of $\triangle PQR$, we can determine the lengths of $\triangle MNO$. Triangle *MNO* is "similar" to triangle *PQR*, which means that the sides are proportional. So the ratio of the smallest sides must equal the ratio of the longest sides. Fill in the numbers and cross-multiply to solve. (Remember Mantra #21, "When you see a proportion, cross-multiply.")

$$\frac{\text{smallest}}{\text{smallest}} = \frac{\text{longest}}{\text{longest}} \Rightarrow \frac{3}{4.5} = \frac{6}{x}$$

So $3x = 27$ and $x = 9$.

Correct answer: H

ACT Math Mantra #26
Similar triangles have sides that are proportional.

Dr. Evil and Mini-Me Drills

Medium

1 In the figure below, △XYZ and △PQR are similar triangles with the side lengths given in feet. What is the perimeter, in feet, of △XYZ ?

Ⓐ 18
Ⓑ 27
Ⓒ 45
Ⓓ 54
Ⓔ 69

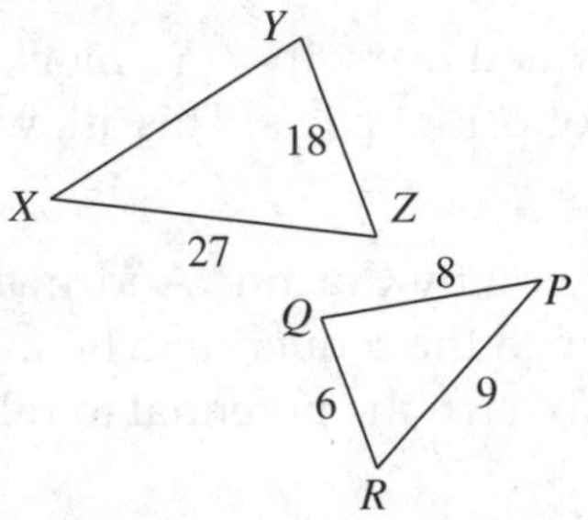

2 Isosceles triangle △ABC is similar to triangle △MNO. If AC = 4, AB = 5, BC = 4, and MN = 15, what is the measure of MO ?

Ⓕ 5
Ⓖ 10
Ⓗ 12
Ⓙ 15
Ⓚ 20

3 A person 2 meters tall casts a shadow 4 meters long. At the same time a lamppost casts a shadow 15 meters long. How many meters tall is the lamppost?

Ⓐ 3.75
Ⓑ 7.5
Ⓒ 15.25
Ⓓ 21.5
Ⓔ 25

4 Triangle △PQR is similar to triangle △XYZ. $\overline{PQ}$ is 6 centimeters, $\overline{QR}$ is 8 centimeters, and $\overline{PR}$ is 14 centimeters. If the longest side of △XYZ is 35 centimeters, what is the perimeter, in centimeters, of △XYZ ?

Ⓕ 63
Ⓖ 70
Ⓗ 83
Ⓙ 105
Ⓚ 175

Hard

5 In the figure below, A is on $\overline{XY}$ and B is on $\overline{ZY}$. What is the length of $\overline{BY}$?

Ⓐ 614.5
Ⓑ 625.5
Ⓒ 1237.5
Ⓓ 1490.5
Ⓔ 2500

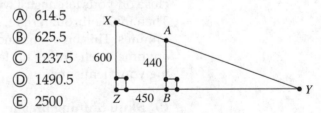

Most kids had trouble with fractions back in 3rd grade. This makes sense, since an 8-year-old's brain is not usually ready to deal with fractions. Nine years later, however, even though a 17-year-old's brain is more than ready, many teenagers still believe that they "suck" at fractions. They are unknowingly holding on to an outdated and unnecessary belief.

Yoga can help you get past pesky outdated fears. It opens blocked pathways in the brain and body, dropping the clouded lens of prior beliefs. This allows you to see clearly and experience reality as it truly is.

Here is a yoga sequence designed for you by Corinne Andrews, a yoga teacher in western Massachusetts. Corinne designed the sequence to help you relieve stress, get energized, and focus your mind. We all have the potential to release fear, trust ourselves, and live freely.

Record these instructions into your iPod. Then find a comfortable spot and get yoga-ing. If you can't say "buttocks" without giggling, visit www.brianleaf.com to download a free podcast of Corinne leading you through the poses.

❶ Alternate Nostril Breathing

Sit in a chair or on the floor with a blanket or pillow underneath your bum. Close off your right nostril with your thumb, and inhale slowly through the left nostril. Then close off your left nostril with your ring finger, and exhale slowly through the right. Then inhale through right, switch fingers, and exhale through left. Repeat this for a few minutes. This breath balances the right and left hemispheres of your brain. You can do it anytime, even during the test. If you get a strange look, just explain that you are balancing your hemispheres.

❷ Skull Shining Breath

Exhale sharply out through your nose as if blowing out a candle, and then let the inhale come in naturally. Repeat for 10 breaths, and then take one relaxed deep breath. Repeat the process two times. This clears your mind—and your sinuses.

❸ Cat Lift and Round

Begin on your hands and knees, so you look like a coffee table. Inhale as you look up, allowing your back to arch down, and then slowly exhale as you look down, rounding your back up. Repeat 10 times. This practice gently warms up the spine and nervous system, and relaxes the upper back and shoulders.

❹ Half Sun Salute

Stand with your feet together. Inhale and lift your arms up, then exhale, bending your knees a bit, and slowly fold over to touch the floor. Inhale as you slide your hands up your legs to come up halfway, and then exhale to fold back down. Finally, inhale to come all the way back up. This practice warms up your whole body and helps you connect to the rhythm of your breath.

⑤ Downward Dog

Begin on all fours, with your hands shoulder-width apart. Press firmly down through all parts of your fingers and hands, and reach your tailbone (bum) into the air. Your body should resemble an upside-down V. Keep your knees a bit bent and reach your bum way up. This pose invigorates your entire body and mind, while also relaxing your shoulders and upper back. Stay in this posture for 3 relaxed breaths. Then slowly lower down.

⑤ Fish Pose

Sit in a chair and gently arch your upper back over the back of the chair. Breathe 3 relaxed breaths. This pose opens and stretches your chest, shoulders, and neck. You can also do this pose by placing a rolled blanket or towel underneath your shoulder blades as you lay on the floor.

⑥ Relaxation Pose

Lie on your back on a carpet or mat. Take a few deep breaths, allowing each exhalation to be a long sigh. Allow your body to relax and to be supported by the floor. Relax as your thoughts pass through your mind. Do not engage with them, just witness.
(You can sample a few minutes of relaxing music here, probably not DMX.)
(After a few minutes . . .)
Begin to deepen the breath. . . . Feel your belly and chest rise and fall with each breath. . . . Wiggle your fingers and toes. Then gently roll to one side and come to a seated position. Notice how you feel. Set the intention to take this feeling into your day, into your relationships, into your schoolwork, into your ACT prep. You've entered The Society For Free Living.

"Is" Means Equals— Translation

I love translation questions. It's like you are translating Spanish to English. On the ACT we are translating English to math and here is our dictionary:

"what number"	means	x
"37 percent"	means	0.37
"of" and "product"	mean	multiply
"less"	means	minus
"sum" or "more than"	means	plus
"quotient"	means	divide
"is"	means	equals
"is to" or "per"	means	divide (ratio) and often a proportion
"twice a number"	means	$2x$
"x increased by 25%"	means	$x + 0.25x$

Simply translate, word for word.

Here's the question from the Pretest:

27. If 5 percent of 20 percent of a number is 24 less than one-quarter of the number, what is the number?

    A. 1   **B.** 5   **C.** 20   **D.** 50   **E.** 100

Solution: First translate. Then use algebra to solve.

$(0.05)(0.20)(n) = 0.25n - 24$	Translate, "one-quarter" means 0.25.
$0.01n = 0.25n - 24$	Multiply $(0.01)(0.20)$.
$24 = 0.24n$	Add 24 to both sides, and subtract $0.01n$ from both sides.
$100 = n$	Divide both sides by 0.24.

Correct answer: E

ACT Math Mantra #27
Translate word problems from English to math.

Let's try it, come along children. . . .

"Is" Means Equals—Translation Drills

Easy

1 What is $\frac{1}{8}$ of 30% of $6000 ?

  Ⓐ 22.5

  Ⓑ 225

  Ⓒ 750

  Ⓓ 1500

  Ⓔ 14,400

2 The speed of one airplane exceeds 3 times the speed of another by 20 mph. If s mph is the speed of the slower plane, which of the following expresses the speed, in miles per hour, of the faster plane?

  Ⓕ $s - 20$

  Ⓖ $3s + 20$

  Ⓗ $3s - 20$

  Ⓙ $3s + 60$

  Ⓚ $s + 60$

3 31 is what percent of 155 ?

  Ⓐ 3%

  Ⓑ 5%

  Ⓒ 20%

  Ⓓ 30%

  Ⓔ 500%

Medium

4 In a survey of 1000 kids, if 60% who saw a movie thought it was funny, and if 10% of the kids who thought it was funny will buy the movie, how many kids who saw the movie and thought it was funny will buy it?

  Ⓕ 700

  Ⓖ 600

  Ⓗ 70

  Ⓙ 60

  Ⓚ 10

5 If 125% of a number is 750, what is 65% of the number?

  Ⓐ 360

  Ⓑ 370

  Ⓒ 390

  Ⓓ 410

  Ⓔ 425

Arithmetic Word No-Problems

"Problems" is a bad name for these; they are definitely no-problems. I know some kids who see a word problem and shut down; "I hate these; I have no chance," they say. So untrue! And anyway, imagine you're a soccer goalie. What if you analyzed every ball that came at you, analyzed whether you had a chance or not. You'd wind up watching a lot of balls go past. Go for every ball, every question. I'm not telling you to spend 7 minutes on one question, just to ask yourself, which Skill can I use for this question, what's the easy way to get it? You'll get more of them right, even ones you thought you had no chance on. Intimidation is the only thing standing in your way. So go for it.

Arithmetic word no-problems are really a type of translation. You just have to translate the words into math. Sometimes they are not as direct as the ones in Skill 27, but once you translate, the math is even easier. And every single ACT has at least 5 of these! That's a lot, and they are easy points, total gimmies, money in the bank.

A huge trick for these questions, and for any question containing fractions, is to translate fractions to decimals. This is amazing; it turns what looks like a tough question into a totally easy question. For some kids, this strategy alone gets them 3 more questions per test!

Here's the question from the Pretest:

28. In Seth's refrigerator he found 2 jars of mustard. He estimated that one was $\frac{1}{3}$ full and the other was $\frac{2}{5}$ full. If he combined the two jars into one, approximately how full would the one combined jar be?

    F. $\frac{1}{3}$ full   **G.** $\frac{3}{5}$ full   **H.** $\frac{2}{3}$ full   **J.** $\frac{9}{10}$ full   **K.** Completely full

Solution: First translate fractions to decimals: $\frac{1}{3} = 0.333$ and $\frac{2}{5} = 0.4$.

Then it's too easy! "**Combine** the jars" tells us to add them together.

So the combined jar is $0.333 + 0.4 = 0.733$ full. Now just convert the fractions in the answer choices, and choose the one closest to 0.7333. Notice that this question asks for the approximate answer, so take the closest one to 0.7333.

F. 0.33 full   **G.** 0.60 full   **H.** 0.666 full   **J.** 0.90 full   **K.** Completely full

Choice H is the closest answer to 0.7333 of a jar of delicious yellow mustard.

Correct answer: H

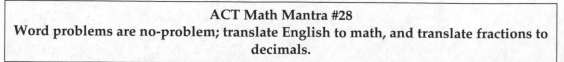

ACT Math Mantra #28
Word problems are no-problem; translate English to math, and translate fractions to decimals.

Let's practice.

Arithmetic Word No-Problems Drills

Easy

❶ What is the total cost, rounded to the nearest penny, of 3.5 pounds of corn at $0.35 per pound and 1.5 pounds of tomatoes at $0.74 per pound?

Ⓐ $1.00
Ⓑ $2.27
Ⓒ $2.34
Ⓓ $3.46
Ⓔ $3.78

❷ Evan pays $65 per year for pay-per-view access and $5 for every 3 movies he watches. Last year, he watched 492 pay-per-view movies. What was the total cost that he paid to pay-per-view last year?

Ⓕ $65
Ⓖ $820
Ⓗ $885
Ⓙ $1473
Ⓚ $1538

❸ Laurie ran $3\frac{4}{5}$ miles on Tuesday and $4\frac{2}{3}$ miles on Wednesday. The total distance, in miles, that Laurie ran during those two days is within which of the following ranges?

Ⓐ At least $7\frac{1}{8}$ and less than $7\frac{4}{5}$
Ⓑ At least $7\frac{2}{5}$ and less than $7\frac{4}{5}$
Ⓒ At least $8\frac{1}{8}$ and less than $8\frac{1}{4}$
Ⓓ At least $8\frac{2}{5}$ and less than $8\frac{2}{3}$
Ⓔ At least $9\frac{1}{8}$ and less than $9\frac{2}{3}$

❹ Alfred bought 3 cases of blueberries. Each case contained twelve 2-pint containers. Alfred could have bought the same amount of berries by buying how many 3-pint containers of berries?

Ⓕ 12
Ⓖ 18
Ⓗ 24
Ⓙ 36
Ⓚ 72

Just Do It!—Springboard

When approaching a question, students often say, "I have no idea where to start!" This book, and this strategy especially, tells you where to start.

> **ACT Math Mantra #29**
> **When something can be factored, foiled, reduced, or simplified—do it.**
> **When you have two equations in a question, ask how they relate.**
> **Convert fractions to decimals.**

Examples:

- When you see $x^2 - y^2$, **factor** it to $(x - y)(x + y)$.
  Note: $x^2 - y^2 = (x - y)(x + y)$ is the ACT's favorite kind of factoring, memorize it!
- When you see $(x - 3)(x + 2)$, **FOIL** it to get $x^2 - x - 6$.
- When you see $\frac{12}{16}$, **reduce** it to $\frac{3}{4}$.

- When you see $\sqrt{50}$, simplify it to $5\sqrt{2}$ or just 7.07.

- When you see $\frac{7}{8}$, use your calculator (top divided by the bottom) to convert it to 0.875.

These steps tell you where to go with a question. I have seen this strategy dramatically help students, especially kids who get stuck and don't know what to do next—this strategy tells you what to do next.

Let's look at the question from the Pretest:

29. If $x^2 - y^2 = 84$ and $x - y = 6$, what is the value of $x + y$?

    A. 6
    B. 8
    C. 10
    D. 12
    E. 14

Solution: When you see $x^2 - y^2$, factor it.

$$x^2 - y^2 = 84$$
$$(x - y)(x + y) = 84 \quad \text{Factor } x^2 - y^2 \text{ to } (x - y)(x + y).$$
$$6(x + y) = 84 \quad \text{Substitute in 6 for the } (x - y), \text{ since } (x - y) = 6.$$
$$(x + y) = 14$$

Correct answer: E

Just Do It!—Springboard Drills

Medium

1 If $x - y = 4$ and $z = 6x - 5 - 6y$, then $z = ?$

(A) -1

(B) 1

(C) 19

(D) 24

(E) 29

2 For $x^2 \neq 49$, $\dfrac{(x-7)^2}{x^2 - 49} = ?$

(F) $\dfrac{x-7}{x+7}$

(G) $\dfrac{1}{x+7}$

(H) $\dfrac{1}{x-7}$

(J) $-\dfrac{1}{7}$

(K) $\dfrac{1}{7}$

3 The expression $\dfrac{8 + \frac{3}{8}}{1 + \frac{3}{16}}$ is equal to which of the following?

(A) 5

(B) 6

(C) $\dfrac{134}{19}$

(D) 8

(E) $\dfrac{255}{16}$

Hard

4 When $\dfrac{a}{b} = 8$, $a^2 - 64b^2 = ?$

(F) 0

(G) 7

(H) 8

(J) -7

(K) -8

Beyond Your Dear Aunt Sally: The Laws of Exponents I

The ACT loves to test the laws of exponents. I know that you think that the ACT is incapable of love, that they are a sadistic organization. But I've visited their building, camped outside, and walked among them, and I can tell you that they are almost normal people.

Anyway, the ACT loves exponents; they are on every test. Memorize these rules. Practice working with them in the drills, and check your answers. Then teach them to a friend.

$n^6 \times n^2 = n^8$ When multiplying matching bases, add exponents.

$\dfrac{n^6}{n^2} = n^4$ When dividing matching bases, subtract exponents.

$(n^6)^2 = n^{12}$ Power to a power, multiply exponents.

$n^0 = 1$ Any base to the 0 power equals 1.

ACT Math Mantra #30
Memorize the laws of exponents.

Let's look at the question from the Pretest:

30. Which of the following expressions is equivalent to $(-2x^2y^2)^3$?

     F. $-2x^5y^5$ **G.** $-8x^6y^6$ **H.** $2x^5y^5$ **J.** $8x^5y^5$ **K.** $8x^6y^6$

Solution: $(-2x^2y^2)^3 = (-2)^3(x^2)^3(y^2)^3 = -8x^6y^6$.

Correct answer: G

Beyond Your Dear Aunt Sally: The Laws of Exponents I Drills

Easy

1 $2x^2 \cdot 3x^3y \cdot 3x^3y$ is equivalent to

(A) $8x^8y^2$

(B) $8x^{18}y^2$

(C) $18x^8y$

(D) $18x^{18}y^2$

(E) $18x^8y^2$

2 What is the product of m^3 and $2m^5$ and $\frac{1}{m^2}$?

(F) $2m^{13}$

(G) $2m^6$

(H) $3m^{13}$

(J) $4m^6$

(K) $2m^{15} - m^2$

3 If x is a real number such that $x^3 = 27$, then $x^3 + x^2 = ?$

(A) 27

(B) 29

(C) 32

(D) 36

(E) 42

Medium

4 Which of the following expressions is equivalent to $(-3x^3y^4)^4$?

(F) $-27x^7y^8$

(G) $-81x^{12}y^{16}$

(H) $27x^7y^8$

(J) $81x^7y^8$

(K) $81x^{12}y^{16}$

5 If it travels at 186,000 miles per second, about how many miles does a ray of light travel in 3 hours?

(A) 2×10^8

(B) 2×10^9

(C) 3×10^{10}

(D) 3×10^{11}

(E) 4×10^{12}

6 If m, n, and p are positive integers such that $m^p = x$ and $n^p = y$, then $xy = ?$

(F) mn^p

(G) mn^{2p}

(H) $(mn)^p$

(J) $(mn)^{2p}$

(K) $(mn)^{p^2}$

Far Beyond Your Dear Aunt Sally: The Laws of Exponents II

I fear that I will always be
A lonely number like root three.
A three is all that's good and right,
Why must my three keep out of sight?
Beneath a vicious square root sign,
I wish instead I were a nine . . .

Kumar Patel, *Harold & Kumar Escape from Guantanamo Bay* (Warner Bros., 2008)

$2n^2 + n^2 = 3n^2$	When adding with matching bases and matching exponents, add coefficients.
$2n + n^2$	Does not combine. When adding, they combine only if they have matching bases and matching coefficients.
$n^{-2} = \dfrac{1}{n^2}$	A negative exponent means "take the reciprocal."
$n^{4/3} = \sqrt[3]{n^4}$	For a fractional exponent, the top number is the power and the bottom number is the root.

Now, let's take a look at the question from the Pretest.

31. If $8m^2p^3 = m^5p$, what is m in terms of p ?

   A. $p^{2/3}$ **B.** $2p^{2/3}$ **C.** $8p^{2/3}$ **D.** $2p^2$ **E.** $8p^{-2}$

Solution: Great review of Skill 3. "What is m in terms of p" means "solve for m—use algebra to get m alone."

$8m^2p^3 = m^5p$	Divide both sides by p.
$8m^2p^2 = m^5$	Divide both sides by m^2.
$8p^2 = m^3$	Put both sides to the one-third exponent, since $(m^3)^{1/3} = m$.
$(8p^2)^{1/3} = m$	Distribute the exponent.
$2p^{2/3} = m$	

Correct answer: B

Far Beyond Your Dear Aunt Sally: The Laws of Exponents II Drills

Easy

① What is the sum of $2x^2$ and $2x^3$?

- (A) $4x^5$
- (B) $2x^6$
- (C) $2x^2 + 2x^3$
- (D) $2x^4 + 2x^4$
- (E) $(2x^2)(2x^3)$

Medium

② The product of $3m^{-2}$ and $2m^{-5}$ is

- (F) $\dfrac{6}{m^7}$
- (G) $\dfrac{1}{6m^7}$
- (H) $5m^7$
- (J) $6m^7$
- (K) $\dfrac{5}{m^7}$

Hard

③ For positive real numbers a, b, and c, which of the following expressions is equivalent to $a^{2/3}b^{1/6}c^{1/2}$?

- (A) $\sqrt{a^2bc}$
- (B) $\sqrt[6]{a^2bc}$
- (C) $\sqrt[6]{a^4bc}$
- (D) $\sqrt[6]{a^4b^3c^3}$
- (E) $\sqrt[6]{a^4bc^3}$

④ If $\left(\dfrac{3}{4}\right)^x = \sqrt{\left(\dfrac{4}{3}\right)^4}$, then $x = ?$

- (F) 2
- (G) 1
- (H) 0
- (J) -1
- (K) -2

The graph of the equation $y = mx + b$ is a line. In the equation, m is the slope, sometimes called the rate of change, and b is the y intercept (the place where the line crosses the y axis). Usually, that is enough to get a question right.

> **ACT Math Mantra #32**
> For the equation $y = mx + b$, m is the slope, and b is the y intercept.

Let's take a look at the question from the Pretest.

32. If Sawyer charges $20 per bottle of water and a flat fee of $25 even to discuss a sale, which of the following equations expresses Sawyer's total fees for x bottles of water?

   F. $y = 45x$
   G. $y = 25x + 20$
   H. $y = 5x$
   J. $y = 20x + 25$
   K. $y = 45x$

Solution: The fee for each bottle, $20, goes next to x. And $25 is a fixed fee, unrelated to how many bottles someone buys, so $25 stands alone: $y = 20x + 25$.

Correct answer: J.

That was so easy to explain that I have some room left. So, here's Brian's Math Magic Trick #2.

> **Brian's Math Magic Trick #2**
> Choose a number between 1 and 10.
> Multiply your number by 9.
> Take the result and add the digits; for example, if your result was 26, then $2 + 6 = 8$.
> Now, take that number and subtract 5.
> Then correspond that number to the alphabet: 1 is A, 2 is B, 3 is C, 4 is D, 5 is E, and so on.
> Choose a country that begins with that letter. Spell the country; what is the second letter?
> Choose an animal that begins with that letter. Now turn to the solutions page.

$y = mx + b$ Drills

Easy

❶ If a movie trailer offends 45 people in a theater, and then the feature movie disturbs 25 more people every scene the villain appears in, which of the following equations expresses the number of people offended or disturbed after x villain scenes in the movie?

Ⓐ $y = 45$

Ⓑ $y = 25x + 45$

Ⓒ $y = 20x$

Ⓓ $y = 70x$

Ⓔ $y = 20x + 70$

Medium

❷ Which of the following equations best fits the information in the table below?

Ⓕ $y = 5x - 2$

Ⓖ $y = -9.5$

Ⓗ $y = 2x - 4$

Ⓙ $y = 0.25x + 3$

Ⓚ $y = -9.1x$

X	Y
4	4
8	5
12	6

❸ Which of the following equations could represent the graph in the standard (x, y) coordinate plane shown below?

Ⓐ $y = 2$

Ⓑ $y = 2x + 2$

Ⓒ $y = 4x$

Ⓓ $y = -x + 2$

Ⓔ $y = 4x + 4$

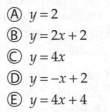

❹ If the cost of advertising on Facebook consists of a fixed charge plus a charge per insertion of the ad, based on the table below, what is the fixed charge?

Ⓕ 1000

Ⓖ 976

Ⓗ 500

Ⓙ 34

Ⓚ 33

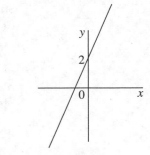

Inserts	Cost
1000	34
2000	35
3000	36

❺ For the first 10 seconds after a geyser erupts, its height is given by $h = 3.4s$, where h is in feet and s represents seconds after erupting. When will the geyser first be over 20 feet tall?

Ⓐ Between 3 and 4 seconds

Ⓑ Between 4 and 5 seconds

Ⓒ Between 5 and 6 seconds

Ⓓ Between 6 and 7 seconds

Ⓔ Between 7 and 8 seconds

Arrangements

Arrangement questions ask you how many arrangements of something are possible, such as how many different ways 4 letters can be arranged. Most tests have at least one of these. Arrangement questions seem impossible to many students, but they are easy if you know the steps:

Step 1. Draw a blank for each position.
Step 2. Fill in the # of possibilities to fill each position.
Step 3. Multiply.

Let's take this strategy for a spin on the question from the Pretest.

33. Arthur Dent buys an ice cream sundae that contains one scoop of ice cream, one sauce, and either a cherry or pineapple wedge on top. He can choose chocolate, vanilla, strawberry, or banana ice cream; he can choose chocolate, caramel, or berry sauce; and he can choose either the cherry or the pineapple wedge for the top. How many different arrangements of these ingredients for Arthur's ice cream sundae are possible?

A. 9 **B.** 14 **C.** 24 **D.** 44 **E.** 64

Solution:

❶ Draw a blank for each position.

$$\underset{\substack{\text{ice cream} \\ \text{flavor}}}{\underline{\hphantom{xxxxxx}}} \times \underset{\substack{\text{syrup} \\ \text{flavor}}}{\underline{\hphantom{xxxxxx}}} \times \underset{\substack{\text{cherry or} \\ \text{pineapple}}}{\underline{\hphantom{xxxxxx}}}$$

❷ Enter the number of possibilities that can fill each position.

$$\underset{\substack{\text{ice cream} \\ \text{flavor}}}{\underline{4\hphantom{xxxx}}} \times \underset{\substack{\text{syrup} \\ \text{flavor}}}{\underline{3\hphantom{xxxx}}} \times \underset{\substack{\text{cherry or} \\ \text{pineapple}}}{\underline{2\hphantom{xxxx}}}$$

❸ Multiply the numbers to get the number of arrangements! $4 \times 3 \times 2 = 24$

Correct answer: C

That's the way to do 95% of ACT arrangement questions. There is only one trick that the ACT tries. If a question is about teams of two or pairs or specifically points out repeated pairs, then divide your answer by 2.

Here's an example: There are 12 children in a class. Two will be chosen as a team to go and hide for a game. How many such teams of two children are possible?

Draw two blanks for the two children on the team. Fill in the number of possibilities for each blank. Multiply the numbers: $12 \times 11 = 132$. Normally you'd be done, but here we have one final step: we divide the answer by 2. For this team of two kids, it doesn't matter if it's Jimmy and Jill or Jill and Jimmy, so there are exactly twice as many answers as there should be. This rule works for any team of two. (If the ACT used a team of three, then we would divide by the number 6, but usually they use teams of two and we divide by the number 2.)

ACT Math Mantra #33
When you see an arrangement question, draw a blank for each position,
fill in the # of possibilities to fill each position, and multiply.
When an arrangement question mentions a "team of two," or specifically
points out repeats, divide your result by 2.

Arrangements Drills

Easy

1 For a fund raiser, the meditation club is selling T-shirts with a choice of two slogans, "See clearly" or "Nonattachment." Each shirt is available in small, medium, or large. How many different types of shirts are available?

(A) 2
(B) 3
(C) 4
(D) 5
(E) 6

2 Five actors are being cast to fill five roles. If each actor plays only one role, how many different arrangements of actors in the five roles are possible?

(F) 5
(G) 10
(H) 60
(J) 120
(K) 240

3 At the build-your-own-burrito bar you can choose chicken, beef, or shrimp. You can include no vegetable, spinach, or sautéed zucchini; and you can top it with mild, medium, hot, or killer salsa. How many different burritos can be ordered?

(A) 8
(B) 18
(C) 24
(D) 36
(E) 48

Medium

4 Kyle will write 3 of the symbols shown below on a banner. How many such arrangements are possible?

(F) 3   ⊗ ♣ ♥ Φ ψ
(G) 5
(H) 15
(J) 30
(K) 60

5 Of the six members of the girls tennis club, two will compete as the doubles team. How many different such teams of two girls are possible?

(A) 6
(B) 12
(C) 15
(D) 30
(E) 60

SohCahToa!

If you were in ancient Greece, the word "trigonometry" would not be intimidating. It would mean "triangle measurement," and when your pal Plato came up and said, "Hey dude, let's do our trigonometry homework," you'd just hear, "Hey dude, let's measure some triangles." But since you don't live in ancient Greece, it sounds like he's saying "Hey dude, let's do our %*#$!^ homework."

But trig is just triangle measurement. It's the way that sides and angles in a triangle are related to each other. They gave these relationships names that made perfect sense in ancient Greece, but now they freak some kids out, so stay relaxed when you see the names. Here they are, and for half the ACT trig questions, they are all you need to know:

$$\sin = \frac{\text{opposite}}{\text{hypotenuse}} \qquad \cos = \frac{\text{adjacent}}{\text{hypotenuse}} \qquad \tan = \frac{\text{opposite}}{\text{adjacent}}$$

This means, for example, that the sin of an angle equals the ratio of the side opposite the angle, divided by the hypotenuse. There's a great way to remember these ratios: SohCahToa. It stands for **S**in = **O**pposite over **H**ypotenuse, **C**os = **A**djacent over **H**ypotenuse, and **T**an = **O**pposite over **A**djacent. I recommend taking a break right now and marching around the room chanting "SohCahToa." You're sure never to forget it. SohCahToa is all you need for half of the trig questions on the ACT.

Remember from Skill 24 that "hypotenuse" is just a fancy word for the longest side of a right triangle; it's always the one opposite the 90° angle. The other two sides, the shorter ones, are called *legs*. And "opposite" or "adjacent" refers to the leg opposite or adjacent to the angle being used.

Let's take a look.

Here's the question from the Pretest.

34. For right triangle $\triangle XYZ$, what is $\tan Z$?

    F. 0.2   **G.** 0.6   **H.** 0.75   **J.** 0.8   **K.** 0.9

Solution: Easy! SohCahToa! The question asks for tan Z. Tan $= \frac{\text{opposite}}{\text{adjacent}}$. That means opposite leg over adjacent leg. So $\tan Z = \frac{\text{opposite}}{\text{adjacent}} = \frac{3}{4} = 0.75$.

Correct answer: H

> **ACT Math Mantra #34**
> **SohCahToa!**

SohCahToa! Drills

Medium

1 Standing on Turtle's back, Vince casts a
4-meter shadow, as shown below. The angle of
elevation from the tip of the shadow to the top
of Vince's head is 40°. To the nearest tenth of a
meter, how high is the top of Vince's head?

Ⓐ 1.6

Ⓑ 3.4

Ⓒ 5.5

Ⓓ 7.2

Ⓔ 9.1

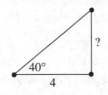

2 If $\sin A = \dfrac{3}{5}$, then which of the following could
be $\tan A$?

Ⓕ $\dfrac{2}{5}$

Ⓖ $\dfrac{3}{4}$

Ⓗ 3

Ⓙ $\dfrac{5}{3}$

Ⓚ 4

3 For right triangle △ABC with dimensions in feet
as shown below, what is $\sin B$?

Ⓐ $\dfrac{5}{7}$

Ⓑ $\dfrac{2\sqrt{6}}{7}$

Ⓒ $\dfrac{2\sqrt{6}}{5}$

Ⓓ $\dfrac{7}{5}$

Ⓔ 12

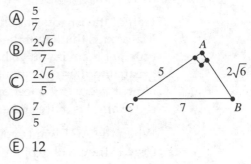

4 For right triangle △XYZ shown below, what is
$\cos Z$?

Ⓕ 0.2

Ⓖ 0.6

Ⓗ 0.75

Ⓙ 0.8

Ⓚ 0.9

Beyond SohCahToa

Half of the trig questions on the ACT are just SohCahToa, which you've now mastered. Many kids think that the rest of the trig questions are way beyond them, that you can only get them if you've had a full-year trig course. If you've had a trig course, that's great, and these questions will be especially easy for you. If not, the exciting news is that almost every trig question can be done with "Use the Answers" or "Make It Real"! You'll see examples of this when we review the Pretest question and in the drills.

When the ACT wants you to use a specific trig concept, such as the *law of sines* or *law of cosines*, they will explain it in the question, and you just have to follow the directions. These are great; they give you the equation, and you just follow the directions. You'll see this also in the drills.

Lastly, occasionally the ACT has a question about the reciprocals of sin, cos, and tan, which are

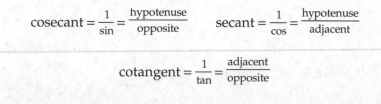

$$\text{cosecant} = \frac{1}{\sin} = \frac{\text{hypotenuse}}{\text{opposite}} \qquad \text{secant} = \frac{1}{\cos} = \frac{\text{hypotenuse}}{\text{adjacent}}$$

$$\text{cotangent} = \frac{1}{\tan} = \frac{\text{adjacent}}{\text{opposite}}$$

Let's apply some of this to the question from the Pretest.

35. What is the value of θ, between 0 and 360, when $\sin \theta = -1$?

    A. 0   **B.** 60   **C.** 135   **D.** 270   **E.** 330

Solution: θ is just the symbol that people use for an angle when it's unknown. (The symbol is called *theta*.) Therefore, "what is the value of θ" means "what is the angle measure?" You probably could have guessed that anyway, since the question tells you that θ is between 0 and 360. Another example of "Don't get intimidated!" Just go with it, make an assumption on the ACT, and you'll probably be correct. They set it up that way. This is a great example of a Beyond SohCahToa question where you really don't need any trig. You just need to "Use the Answers"! Using your calculator, try each choice for θ and see which one gives you $\sin \theta = -1$. Choice D is correct, because $\sin 270 = -1$. If using the "sin" button is new for you, practice it right now. It's easy. Just hit "sin" and type "270" and hit Enter and you'll get an answer of -1.

Correct answer: D

ACT Math Mantra #35
When trig seems tough, "Use the Answers" or "Make It Real."

Beyond SohCahToa Drills

Hard

1 What are the values of θ, between 0° and 360°, when $\tan \theta = 1$?

(A) 45°, 135°, 225°, and 315°

(B) 45° and 135° only

(C) 45° and 225° only

(D) 45° and 315° only

(E) 135° and 315° only

2 Which of the following expressions gives the perimeter of the triangle shown below, with measurements as marked?
(Note: The law of sines states that the ratios between the length of the side opposite any angle and the sine of that angle are equal for all interior angles in the same triangle.)

(F) $59 + \dfrac{32 \sin 52}{\sin 74}$

(G) $59 + \dfrac{27 \sin 52}{\sin 74}$

(H) $59 + \dfrac{32 \sin 78}{\sin 54}$

(J) $59 + \dfrac{32 \sin 78}{\sin 52}$

(K) $\dfrac{32 \sin 52}{\sin 74}$

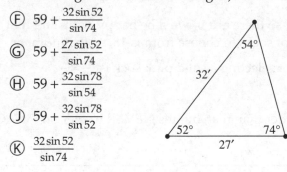

3 Which of the following equations reflects the graph shown below?

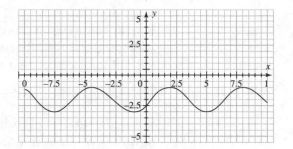

(A) $y = \sin x$

(B) $y = 5 \sin (x - 2)$

(C) $y = 2 \cos x - 2$

(D) $y = \cos (x - 2) - 2$

(E) $y = 2 \tan x - 5$

4 If b, c, and d represent positive real numbers, what is the minimum value of the function $f(x) = \sin b(x - c) - d$?

(F) 0

(G) 1

(H) $-b$

(J) $d + c$

(K) $-1 - d$

Tell Me What You Want, What You Really Really Want . . . Probability

They don't take the ACT over there in England, but Scary, Baby, Ginger, Posh, and Sporty summed up probability in 1996. Like functions, probability could be the topic of a college course, but the ACT asks only one thing! To find out about that one thing, please get out your credit card and see my website.

I'm kidding. The one thing is: $\text{Probability} = \frac{\text{want}}{\text{total}}$.

In English that's "probability equals what you want, divided by the total number of things you are choosing from." Thanks again to Scary, Baby, and the gang.

Let's see that on the Pretest.

36. Of the 18 socks in a drawer, 10 are solid blue, 4 are solid pink, and 4 are pink and blue-striped. If Cherng randomly chooses a sock from the drawer, what is the probability that it will NOT be solid pink?

    F. $\frac{1}{6}$ **G.** $\frac{2}{9}$ **H.** $\frac{4}{9}$ **J.** $\frac{5}{9}$ **K.** $\frac{7}{9}$

Solution: Anytime you see the word "probability," use the equation

$$\text{Probability} = \frac{\text{want}}{\text{total}}$$

We want anything but solid pink socks, so there are 14 we'd be happy with, and that number goes on top. The total number of socks to choose from is 18, so that number goes on bottom. So the probability of not selecting a solid pink sock is $\frac{14}{18} = \frac{7}{9}$.

Correct answer: K

Probability in school could be much harder than that, but it's NEVER harder on the ACT. $\text{Probability} = \frac{\text{want}}{\text{total}}$ is all they ask!

ACT Math Mantra #36
When you see the word "probability," use the equation $\text{Probability} = \frac{\text{want}}{\text{total}}$.

Tell Me What You Want, What You Really Really Want . . . Probability Drills

Easy

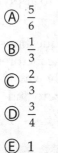

1. If 8 out of 12 marbles in a bag are green, what is the probability that a marble selected at random from the bag will NOT be green?

   Ⓐ $\frac{5}{6}$

   Ⓑ $\frac{1}{3}$

   Ⓒ $\frac{2}{3}$

   Ⓓ $\frac{3}{4}$

   Ⓔ 1

③ The triangle shown below is used at a carnival. People try to toss a penny onto the shaded region to win a prize. If the area of the whole figure is 12 and the area of the unshaded triangle is 7.5, what is the probability of randomly tossing a penny onto the shaded region?

   Ⓐ 4.5

   Ⓑ 1

   Ⓒ $\frac{15}{24}$

   Ⓓ $\frac{9}{24}$

   Ⓔ $\frac{7.5}{12}$

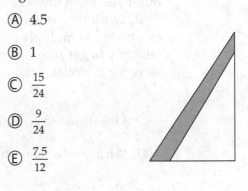

Medium

2. A bag contains 30 marbles, all solid-colored. Each marble is blue, red, or yellow. If the probability of choosing a blue marble is $\frac{2}{3}$ and the probability of choosing a red marble is $\frac{1}{6}$, how many yellow marbles are in the bag?

   Ⓕ 5

   Ⓖ 6

   Ⓗ 7

   Ⓙ 8

   Ⓚ 9

Anything Times Zero Is Zero

"Anything Times Zero Is Zero" questions ask you to name the solutions to an equation like $(x - 3)(x + 2) = 0$. For 90% of these questions, you can just "Use the Answers." You don't need any new info or knowledge, just "Use the Answers"! The goal, then, of this drill section is just to get you accustomed to the wording of the questions and to applying "Use the Answers."

But if you want the behind the scenes, here's the "math class way" for these questions. First, factor the polynomial if it's not already factored. For the two parts of the factored expression to multiply to get zero, one of them must equal zero, since the only way to multiply **to get** zero is to multiply **by** zero. So, set each parenthesis equal to zero and solve for x. Voila!

Let's see these two methods on the question from the Pretest.

37. What are the values for x that satisfy the equation $(x + 4)(x - 3) = 0$?

    A. −4 and 4
    B. −3 and 3
    C. −12
    D. 4 and −3
    E. −4 and 3

Solution: Just "Use the Answers"! And use the process of elimination. If −4 does not work, you can cross out choice A without even trying +4. Find the one answer choice with values that work in the equation. Or do the algebra. For the two parts $(x + 4)$ and $(x - 3)$ to multiply to get zero, one of them must equal zero, since the only way to multiply **to get** zero is to multiply **by** zero. So $(x + 4) = 0$ or $(x - 3) = 0$. Solve for both possibilities:

$(x + 4) = 0$ subtract 4 from both sides. $(x - 3) = 0$ add 3 to both sides.

$x = -4$ $x = 3$

Correct answer: E

ACT Math Mantra #37
For questions like $(x + 4)(x - 3) = 0$, just "Use the Answers" or set each parenthesis equal to zero and solve for x.

Anything Times Zero Is Zero Drills

Easy

1 What are the values for x that satisfy the equation $(x - 2)(x + 5) = 0$?

Ⓐ -2 and 5

Ⓑ -3 and 5

Ⓒ -15

Ⓓ 2 and -5

Ⓔ -2 and 3

2 What are the values for x that satisfy the equation $(2x + 6)(2x - 3) = 0$?

Ⓕ 3 and 3

Ⓖ 6 and 2

Ⓗ -12

Ⓙ -6 and $-\dfrac{3}{2}$

Ⓚ -3 and $\dfrac{3}{2}$

Medium

3 What are the values for x that satisfy the equation $x^2 + 4x + 3 = 0$?

Ⓐ -1 and 4

Ⓑ -1 and -3

Ⓒ 4

Ⓓ 4 and -3

Ⓔ -4 and 5

4 What is the sum of the 2 solutions of the equation $x^2 + 3x - 10 = 0$?

Ⓕ -5

Ⓖ -3

Ⓗ 2

Ⓙ 5

Ⓚ 7

Hard

5 Which of the following expressions is NOT a polynomial factor of $a^4 - 9$?

Ⓐ $a^2 - 3$

Ⓑ $a^2 + 3$

Ⓒ $a - \sqrt{3}$

Ⓓ $a + \sqrt{3}$

Ⓔ $a - 3$

$$Y = ax^2 + bx + c$$

When x is squared (x^2) in an equation, the graph forms a U shaped curve called a *parabola*. The ACT usually shows the equation in standard form, $y = ax^2 + bx + c$, or vertex form, $y = a(x - h)^2 + k$. Either equation is called *quadratic*, which is just fancy vocab for having an x^2 term.

Standard Form

In the equation $y = ax^2 + bx + c$, the a tells whether the U-shaped graph opens up or down, and the c is the y intercept, the place where the graph crosses the y axis.

Vertex Form

In the second equation $y = a(x - h)^2 + k$, the h and k give the coordinates of the vertex of the graph (h,k), and the a again tells whether the U-shaped graph opens up or down. The vertex is the highest or lowest point of the graph and is therefore also called the maximum or minimum point.

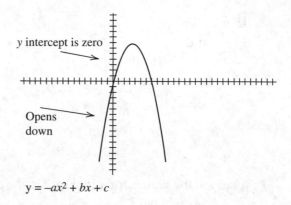

y intercept is zero

Opens down

$y = -ax^2 + bx + c$

This equation can also be written $y - k = a(x - h)^2$, by just subtracting the k to the other side of the equation.

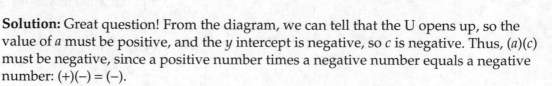

Let's use this on the question from the Pretest.

38. If the graph of $y = ax^2 + bx + c$ is shown below, then the value of ac can be

    F. positive only
    G. negative only
    H. zero only
    J. positive or negative
    K. positive, negative, or zero

Solution: Great question! From the diagram, we can tell that the U opens up, so the value of a must be positive, and the y intercept is negative, so c is negative. Thus, $(a)(c)$ must be negative, since a positive number times a negative number equals a negative number: $(+)(-) = (-)$.

Correct answer: G

ACT Math Mantra #38

For the equation $y = ax^2 + bx + c$, the a tells whether the U-shaped graph opens up or down, and the c is the y intercept. For the equation $y = (x - h)^2 + k$, the h and k give the coordinates of the vertex of the graph (h, k). The vertex is the highest or lowest point of the graph and is therefore also called the maximum or minimum point.

$y = ax^2 + bx + c$ Drills

Medium

1 What are the coordinates for the y intercept in the graph of $y = x^2 + 2x - 3$?

Ⓐ $(3, 0)$
Ⓑ $(-3, 0)$
Ⓒ $(0, 3)$
Ⓓ $(0, -3)$
Ⓔ $(0, 0)$

2 What are the coordinates for the maximum point in the graph of $y - 2 = -(x + 3)^2$?

Ⓕ $(3, 2)$
Ⓖ $(-3, 2)$
Ⓗ $(3, -2)$
Ⓙ $(-3, -2)$
Ⓚ $(0, 0)$

3 The graph of $y = ax^2 + bx + c$ is shown below. When $y = 0$, x has

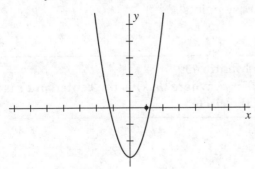

Ⓐ 2 positive solutions
Ⓑ 2 negative solutions
Ⓒ 1 positive solution
Ⓓ 1 negative solution
Ⓔ 1 positive and 1 negative solution

Hard

4 The graph of one of the following equations is the parabola shown below. Which one could it be?

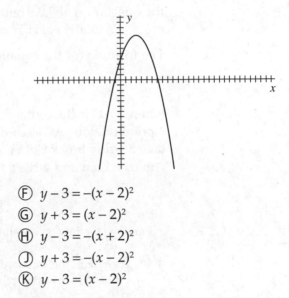

Ⓕ $y - 3 = -(x - 2)^2$
Ⓖ $y + 3 = (x - 2)^2$
Ⓗ $y - 3 = -(x + 2)^2$
Ⓙ $y + 3 = -(x - 2)^2$
Ⓚ $y - 3 = (x - 2)^2$

5 If $y = kx^2 + 3x + r$ is the equation for the parabola shown below, then the product of k and r is

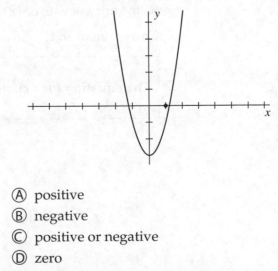

Ⓐ positive
Ⓑ negative
Ⓒ positive or negative
Ⓓ zero
Ⓔ undefined

Circles

I love this Skill. Yes, I do. And I'll tell you why. Almost every ACT has one question asking you to determine the equation of a circle. Nobody has this thing memorized, so the question is always ranked as a "medium," or even a "hard." But if you know the formula, it's totally easy! These are the best.

The formula for the equation of a circle is

$$(x - h)^2 + (y - k)^2 = r^2$$

where (h, k) is the center of the circle and r is the radius. Memorize this and you will gain a point on your ACT score, guaranteed. Don't you love this book! In fact, I want you to take a study break right now to call two friends and recommend this book. Or, log onto Amazon.com and write a rave review expressing your gratitude and appreciation!

Let's see the Pretest question.

39. A circle in the standard (x, y) coordinate plane has center $(4, 2)$ and radius 5 coordinate units. Which of the following is an equation of the circle?

  A. $(x - 4)^2 - (y - 2)^2 = 5$
  B. $(x + 4)^2 + (y + 2)^2 = 5$
  C. $(x - 4)^2 + (y - 2)^2 = 5$
  D. $(x - 4)^2 + (y + 2)^2 = 25$
  E. $(x - 4)^2 + (y - 2)^2 = 25$

Solution: The center is $(4, 2)$, so $h = 4$ and $k = 2$. Plus, the radius is 5, so $r = 5$. Easy. Plug that into the equation for a circle, $(x - h)^2 + (y - k)^2 = r^2$, which you have memorized. So the answer is E, or $(x - 4)^2 + (y - 2)^2 = 25$. Easy points!

Correct answer: E

ACT Math Mantra #39
The equation for a circle is $(x - h)^2 + (y - k)^2 = r^2$, where (h, k) is the center and r is the radius of the circle.

Circle Drills

Medium

1 Which of the following best describes all points in a plane that are 5 inches from a given point in the plane?

- (A) A circle with a 5 inch radius
- (B) A circle with a 5 inch diameter
- (C) A circle with a 25 inch radius
- (D) A rectangle with 5 inch sides
- (E) A sphere with a 5 inch diameter

2 What is the center of the circle given by the equation $(x - 8)^2 + (y + 2)^2 = 49$?

- (F) $(8, 2)$
- (G) $(7, 49)$
- (H) $(8, -2)$
- (J) $(7, -2)$
- (K) $(8, -7)$

3 A circle in the standard (x, y) coordinate plane has center $(5, -3)$ and radius 6 coordinate units. Which of the following is an equation of the circle?

- (A) $(x - 5)^2 - (y - 3)^2 = 6$
- (B) $(x + 5)^2 + (y + 3)^2 = 6$
- (C) $(x - 5)^2 + (y - 3)^2 = 6$
- (D) $(x - 5)^2 + (y + 3)^2 = 36$
- (E) $(x + 5)^2 - (y + 3)^2 = 36$

Hard

4 Which of the following is an equation of a circle with center at $(1, 4)$ and tangent to the y axis in the standard (x, y) coordinate plane?

- (F) $x^2 + y^2 = 1$
- (G) $x^2 - y^2 = 1$
- (H) $(x - 1)^2 + (y - 4)^2 = 2$
- (J) $(x - 1)^2 + (y - 4)^2 = 1$
- (K) $(x + 1)^2 + (y + 4)^2 = 1$

5 A circle in the standard (x, y) coordinate plane is tangent to the x axis at 3 and tangent to the y axis at 3. Which of the following is an equation of the circle?

- (A) $x^2 + y^2 = 3$
- (B) $(x - 3)^2 - (y - 3)^2 = 9$
- (C) $(x - 3)^2 + (y - 3)^2 = 3$
- (D) $(x - 3)^2 + (y - 3)^2 = 9$
- (E) $(x + 3)^2 + (y + 3)^2 = 9$

Weird Circle Factoid

There's this one weird circle thing that for some reason the ACT loves to use. Know it and you'll gain points. Here it is.

A triangle formed by three points of a circle, where one of the sides of the triangle is a diameter of the circle, is always a right triangle, and the right angle is opposite the diameter. I know, that sounded complicated, but look at the diagram to the right, and it'll make total sense. Look it over as long as you need to, cause there's a good chance that this very factoid will be tested on your ACT.

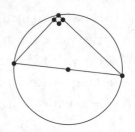

Everybody loves a good circle factoid. Or maybe it's just me since I have a large round head (true) and like circles (true).

Let's take a look at the Pretest question.

40. In the figure below, segment NO is a diameter of the circle, M is a point on the circle, and $MN = MO$. What is the degree measure of $\angle MNO$?

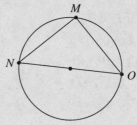

   F.   30
   G.   45
   H.   60
   J.   90
   K.   Cannot be determined from the given information

Solution: Great geometry review! Since triangle $\triangle NMO$ has one side that is a diameter of the circle, the angle opposite to that side is a right angle. So angle $\angle NMO = 90$. Then since $NM = MO$, the other two angles of the triangle must be equal, and $180 - 90 = 90/2 = 45$ each. So $\angle MNO = 45$. This is a great geometry review of Skill 5 (180 degrees in a triangle) and Skill 7 (When you see a triangle with two equal sides, mark the two opposite angles as equal).

Correct answer: G

ACT Math Mantra #40
**If one side of a triangle is the diameter of a circle, and the opposite vertex
is on the circle, then the triangle is right, with its right angle opposite
the diameter.**

Let's drill circles in general, and then our fun new circle factoid.

Weird Circle Factoid Drills

Easy

1 What is the area of a circle with a diameter of 20 inches?

   Ⓐ 10

   Ⓑ 100

   Ⓒ 10π

   Ⓓ 20π

   Ⓔ 100π

Medium

2 In the circle shown below, with 6-cm radius, if angle ∠ACB = 30°, what is the length of arc *AB* ?

   Ⓕ 1

   Ⓖ π

   Ⓗ 6π

   Ⓙ 30

   Ⓚ 12π

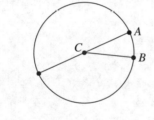

3 In the circle shown below, with 6-cm radius, if angle ∠ACB = 30°, what is the **area** of sector *ABC* ?

   Ⓐ π

   Ⓑ 2π

   Ⓒ 3π

   Ⓓ 12π

   Ⓔ 36π

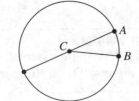

4 The ratio of the radii of two circles is 2:3. What is the ratio of their circumferences?

   Ⓕ 2:3

   Ⓖ 4:9

   Ⓗ 8:27

   Ⓙ 2:π

   Ⓚ π:3

Hard

5 A 3-cm by 4-cm rectangle is inscribed in a circle as shown below. What is the circumference of the circle?

   Ⓐ 5

   Ⓑ 10

   Ⓒ 5π

   Ⓓ 10π

   Ⓔ 25π

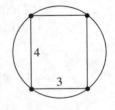

Absolute Value

$|-3|$ means "the absolute value of -3." The $|\ |$ bars mean absolute value. Like many ACT topics, this is really just vocab. If you don't know the meaning of the term, then it's very hard or impossible to get the question correct; but if you know the vocab, it's EASY!

Basically, absolute value just means, "Ditch the negative sign!" $|-3| = 3$, and $|3| = 3$. So it does not mean switch the sign necessarily, it just means drop the $(-)$ negative signs, which makes it positive. You only do this after you've done what is between the bars. For example, $|-3 - 6| = |-9| = 9$. You don't just drop negative signs right away; first you do the math between the bars, and when you've done all the math that you can, then you drop the negative sign.

The key to absolute value on the ACT is "Use the Answers." Absolute value questions are almost always designed so you can just try the answer choices in the question and see which one works. The questions where you cannot "Use the Answers" have variables in the question and variables in the answer. These are very theoretical. Do we just give up and move on? No sir, we "Make It Real." Just choose numbers to try for the variables and see which answer choices work. We'll practice this in the drills.

That's it. That's absolute value. If you did not know this and you do by the end of the drills, then you will gain points!

Let's practice.

41. Which of the following are solutions to $|n + 3| = 5$?

    I. 2   II. -2 III. -8

    A. I only   **B.** III only   **C.** II and III   **D.** I and III   **E.** I, II, and III

Solution: Great "Use the Answers" review. Try each answer choice in the absolute value equation in the question:

    I. $|2 + 3| = 5$—correct   II. $|-2 + 3| \neq 5$—incorrect   III. $|-8 + 3| = 5$—correct

So I and III are correct and D is the answer.

Correct answer: D

ACT Math Mantra #41
When you see absolute value on the ACT, "Use the Answers" or "Make It Real," and remember that absolute value means "Ditch the negative sign."

Absolute Value Drills

Easy

1 If $|4 - m| = 12$, then $m = ?$

   Ⓐ 8 or 0

   Ⓑ 16 or −8

   Ⓒ −16 or −8

   Ⓓ 16 or 8

   Ⓔ 0 or −8

Medium

2 The temperature t in a certain country in spring is modeled by the inequality $|t - 12| \leq 20$. Which of the following is not in that range?

   Ⓕ 32

   Ⓖ 16

   Ⓗ −4

   Ⓙ −6

   Ⓚ −16

3 If $|1 + 2x| < 10$, which of the following is a possible value of x ?

   Ⓐ −6

   Ⓑ −5

   Ⓒ 5

   Ⓓ 6

   Ⓔ 10

4 If $|x| = x + 8$, then $x = ?$

   Ⓕ −6

   Ⓖ −4

   Ⓗ −2

   Ⓙ 6

   Ⓚ 10

Hard

5 Which of the following expressions, if any, are equal for all real numbers m ?

   I. $\sqrt{m^2}$ II. $|-m|$ III. $-|m|$

   Ⓐ I and II only

   Ⓑ I and III only

   Ⓒ II and III only

   Ⓓ I, II, and III

   Ⓔ None of the expressions are equivalent.

Sequences

I love to teach these. Many kids don't know much about sequences, but they are super easy to learn and easy to get right once you know the Skill. They are easy points to add to your ACT score.

The ACT uses two vocab words that you need to know:

❶ **Arithmetic sequence**—a sequence of numbers where a certain number is added to each term to arrive at the next, like 3, 7, 11, 15, 19, . . .
The number 4 is added to each term to arrive at the next.

❷ **Geometric sequence**—a sequence of numbers where a certain number is multiplied by each term to arrive at the next, like 3, −6, 12, −24, 48, . . .
The number −2 is multiplied by each term to get the next.

The ACT asks you to do two things with these. They ask you to either predict the next term or predict the sum of a bunch of terms. There is a complex formula to do this that you may or may not have used in school, but on the ACT we don't need it. Just figure out the number being added or multiplied and write out as many terms as you need to answer the question. We'll do this on the drills.

ACT Math Mantra #42
An arithmetic sequence is a sequence of numbers where a certain number is
ADDED to each term to arrive at the next, like 3, 7, 11, 15, 19.
A geometric sequence is a sequence of numbers where a certain number is
MULTIPLIED by each term to arrive at the next, like 3, −6, 12, −24, 48.

Here's the question from the Pretest.

42. Which of the following is NOT true about the arithmetic sequence 20, 13, 6, −1, . . . ?

    F.   The fifth term is −8.
    G.   The sum of the first five terms is 30.
    H.   The seventh term is −22.
    J.   The common difference of terms is −7.
    K.   The common ratio of consecutive terms is −7.

Solution: When you see an arithmetic sequence, figure out the number being added to each term to get the next. Usually you can tell just by looking at it, or you can subtract: $13 - 20 = -7$. So −7 is being added to each term to arrive at the next. You can even test to make sure that you are correct. Yup, $13 + (-7) = 6$. So now we can predict the fifth and eighth terms, and we can calculate the sum of the first five terms. Choices F, G, H, and J are all correct. K is not true; −7 is the difference, not the ratio. Ratio would be for a geometric sequence.

Correct answer: K

Sequences Drills

Easy

1 What 2 numbers should be placed in the blanks so that the difference between consecutive numbers is the same?

$$14, \underline{\quad}, \underline{\quad}, 47$$

(A) 11, 11
(B) 25, 36
(C) 27, 38
(D) 29, 39
(E) 30, 30

2 The second term of an arithmetic sequence is −5, and the third term is 12. What is the first term?

(F) −22
(G) −17
(H) 0
(J) 17
(K) 22

3 What is the next term after $-\frac{1}{2}$ in the geometric sequence $4, -2, 1, -\frac{1}{2}, \ldots$?

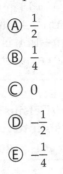

(A) $\frac{1}{2}$
(B) $\frac{1}{4}$
(C) 0
(D) $-\frac{1}{2}$
(E) $-\frac{1}{4}$

Hard

4 Tiara earned 12 dollars on her first day selling lemonade, and she was determined to earn 3 more dollars each day than she had made the day before. If she met this goal exactly, how much money did she earn, in total, for her first 20 days selling lemonade?

(F) 810
(G) 610
(H) 405
(J) 72
(K) 60

It's Gettin' Hot in Here . . .
Fahrenheit/Celsius Conversions

I think that Nelly would agree that it does not get sweeter than these Fahrenheit/Celsius conversion questions. Almost every test has one. Master this page and the drills that follow, and that question is guaranteed points. It never varies, no surprises, it always looks the same. I love questions that we can totally predict!

ACT Math Mantra #43

For a Fahrenheit/Celsius conversion question, when you are given degrees Celsius, just plug in and simplify; but when you are given degrees Fahrenheit, you can either do the algebra or "Use the Answers."

Let's take a look at the question from the Pretest.

43. The temperature 357°F is the point at which mercury will boil. Since Fahrenheit and Celsius temperatures are related by the formula $F = \frac{9}{5}C + 32$, to the nearest degree, which of the following is the boiling point of mercury in degrees Celsius?

    A. 11°C
    B. 87°C
    C. 112°C
    D. 166°C
    E. 181°C

Solution: For a Fahrenheit/Celsius question, you can either do the algebra or just "Use the Answers," whichever you prefer. To "Use the Answers," plug in 357 for F, and then try each answer for C, to see which choice works. To do the algebra, plug in 357 for F and solve for C.

$F = \frac{9}{5}C + 32$

$357 = \frac{9}{5}C + 32$ subtract 32 from each side

$325 = \frac{9}{5}C$ multiply both sides by $\frac{5}{9}$

$180.55 = C \approx 181$

Correct answer: E

Fahrenheit/Celsius Conversions Drills

Easy

1 The relationship between temperature expressed in degrees Fahrenheit (F) and degrees Celsius (C) is given by the formula $F = \frac{9}{5}C + 32$. Winson reads a temperature of 34° on a Celsius thermometer. To the nearest degree, what is the equivalent temperature on a Fahrenheit thermometer?

(A) 7°F

(B) 51°F

(C) 86°F

(D) 93°F

(E) 100°F

Medium

2 A Fahrenheit thermometer reads 63°F. If the temperature increases 10°F, to the nearest degree, what is the new temperature C, in degrees Celsius?

(Note: $F = \frac{9}{5}C + 32$)

(F) 17°C

(G) 23°C

(H) 34°C

(J) 73°C

(K) 77°C

3 The relationship between temperature expressed in degrees Fahrenheit (F) and degrees Celsius (C) is given by the formula $F = \frac{9}{5}C + 32$. If the temperature is 5 degrees Fahrenheit, what is it in degrees Celsius?

(A) −23°C

(B) −15°C

(C) −5°C

(D) 12°C

(E) 41°C

On average, after memorizing and practicing Skills 44 and 45 for avoiding careless errors, most students gain 2 points on their ACT Math score! So learn these strategies.

Most Common ACT Math Careless Errors

❶ Practice being focused, yet relaxed. You don't need to be tense and a wreck to excel on the ACT or in life. You can be intense and work quickly, yet be relaxed. In fact, being intent and focused, yet calm and clear will help you avoid errors. Be fully present with each question: focused, relaxed, awake, and mindful. See "Yoga," page 66.

❷ $(2x)^2 = 4x^2$, not $2x^2$.
Square both the 2 and the x.

❸ $-2(3x - 3) = -6x + 6$, not $-6x - 6$.
Remember to distribute the -2 to both the $3x$ and the -3.

❹ $\frac{5x + 20}{5} = x + 4$, not $x + 20$.
Remember that the 5 is under not only the $5x$, but also the 20.

Let's look at the Pretest question.

44. If $f(x) = 3x^2$, which of the following expresses $f(2p)$?

    F. $6p$
    G. $6p^2$
    H. $12p$
    J. $12p^2$
    K. $24p^3$

Solution: Nice functions review! Remember the key to functions on the ACT—whatever is in the parentheses gets plugged in for x. So, $f(2p)$ just means plug in $2p$ for x. So $f(2p) = 3(2p)^2 = 3(4p^2) = 12p^2$. Careless Error Busters: Remember that the 2 gets squared. Also remember order of operations, multiply 3 times 4 after squaring 2.

Correct answer: J

ACT Math Mantra #44
**"Careless errors are bad, mmmkay," so underline all vocabulary words
and remember to distribute the negative.**

Don't Even Think About It Drills

Medium

1 $(9x - 3) - (3x + 6)$ is equivalent to

- (A) $3(x + 3)$
- (B) $3(2x + 3)$
- (C) $3(2x - 3)$
- (D) $3(3x - 1)$
- (E) $3(x - 3)$

2 If $f(x) = -2x^3$, which of the following expresses $f(-p)$?

- (F) $-2p$
- (G) $2p^3$
- (H) $-2p^3$
- (J) $8p^3$
- (K) $-8p^3$

3 When $m = -1$, which of the following is equivalent to $m(2x^2 - 2)$?

- (A) $-2x^2 + 2$
- (B) $-2x^2 - 2$
- (C) $2x^2 - 2$
- (D) $2x^2 + 2$
- (E) $x^2 + 2$

4 If, $y = \dfrac{3a - 15}{x}$, find y when $x = 3$.

- (F) $a - 15$
- (G) $9a - 15$
- (H) $9a - 5$
- (J) $a - 5$
- (K) $a - 3$

Hard

5 If $m = 2p$, then which of the following is equivalent to $(m + 4)^2$?

- (A) $2p + 4$
- (B) $4p^2 + 4$
- (C) $2p^2 + 8p + 16$
- (D) $4p^2 + 8p + 16$
- (E) $4p^2 + 16p + 16$

5 If $f(x) = x(2x^2 - 2)$, find $f(-2p)$.

- (F) $-16p^3 + 4p$
- (G) $-8p^3 + 4p$
- (H) $-8p^3 - 4p$
- (J) $16p^3 - 4p$
- (K) $8p^3 + 4p$

Don't Even Think About It! . . . Most Common Careless Errors II

"Careless errors are bad, mmmkay." So, let's learn to avoid five more of the most common ones.

⑤ $(x + 3)^2 = (x + 3)(x + 3) = x^2 + 6x + 9$, not $x^2 + 9$.
Remember the middle term when you FOIL.

⑥ $2(3)^2 = 18$, not 36.
Remember your old friend PEMDAS, order of operation.
Parenthesis, Exponents, Multiplication/Division, Addition/Subtraction

⑦ Remember to finish the question. Sometimes you have gotten a number for **part** of a question and the ACT provides that number as an answer choice. So before you grid each answer, take a moment to pause and say, "Wait (insert your name here), did I **finish** the question?" This is so simple, but hugely important. This is the most common careless error, and it's so easy to avoid!

⑧ $(x^2)(4x^5)(3x) = 12x^8$ not $12x^7$.
Remember that x without an exponent means x^1.

⑨ Lots of ACT questions mention units, like feet or centimeters. And 90% of these are just there to make the question realistic and are not at all tricky. But when a question has feet and inches or hours and minutes in the same question, remember to convert correctly. For example, 3.5 feet is not 3 feet 5 inches; it is 3 feet 6 inches. And 2.5 hours is not 2 hours 50 minutes; it is 2 hours 30 minutes. Of course you know this, but it's easy to forget when rushing to solve a question. As long as you know to keep an eye out for it, you'll spot it when it comes up.

Students lose way too many points on careless errors. Don't do that. Memorize the above rules and practice watching for them. Be focused, relaxed, and mindful, and you will not make careless errors. You will be a math ninja, and like most ninja of ancient Japan, you will ace the ACT.

Pretest question:

45. If a board 9 feet 10 inches long is cut in half, how long is each new piece?

    A. 4'9"   **B.** 4'11"   **C.** 5"   **D.** 5'2"   **E.** 5'5"

Solution: Great question. So easy to make a careless error, unless you know to watch for it. To divide 9 feet 10 inches by 2, let's do feet and inches separately. Here's the key, 9 feet divided by 2 is 4' 6", not 4' 5", and 10 inches divided by 2 is 5". So 4'6" plus 5" equals 4'11". Nasty tricksy Hobbitses.

Correct answer: B

Don't Even Think About It! . . . Drills

Medium

1 If $f(x) = 3x^2$, which of the following expresses $f(4p)$?

- Ⓐ $12p^2$
- Ⓑ $32p^2$
- Ⓒ $48p^2$
- Ⓓ $64p^2$
- Ⓔ $144p^2$

2 In a timed run, Maiko posted a time of 1 minute 45 seconds for a quarter-mile course. If she could maintain that pace, how many minutes would it take her to run 1 mile?

- Ⓕ 8.2
- Ⓖ 7.0
- Ⓗ 6.4
- Ⓙ 5.8
- Ⓚ 2.9

Hard

3 If $x = 3y$, then which of the following is equivalent to $(x - 2)^2$?

- Ⓐ $3y - 2$
- Ⓑ $3y^2 - 4$
- Ⓒ $3y^2 + 6y - 4$
- Ⓓ $9y^2 - 12y + 4$
- Ⓔ $9y^2 + 18y + 4$

4 For functions p and r such that $p(x) = 2x(x - 2)$ and $r(x) = \frac{x}{3}$, find $r(p(-4))$.

- Ⓕ 0.5
- Ⓖ 10
- Ⓗ 12
- Ⓙ 14
- Ⓚ 16

5 Josh rode his bicycle from his house to his grandmother's house at a rate of 10 miles per hour. Later that day he rode home at a rate of 15 miles per hour. If the round trip took a total of 30 minutes, how many miles does Josh live from his grandmother?

- Ⓐ 0.3
- Ⓑ 3.0
- Ⓒ 4.0
- Ⓓ 4.5
- Ⓔ 8

Misbehaving Numbers: Weird Number Behavior

The ACT tests your understanding of how certain weird numbers behave, such as when you subtract a negative, it's like adding; –6 is smaller than –1; and one-half squared gets smaller. You can know these too, because here's the thing, the ACT always tests the same weird math behaviors. Here they are:

❶ Small fractions multiplied by small fractions get smaller.
Usually when you multiply numbers, the result is larger than the originals, but small fractions get smaller.

Example: $\left(\frac{1}{2}\right)^2 = \frac{1}{2} \times \frac{1}{2} = \frac{1}{4}$

❷ The larger the digits of a negative number, the smaller it actually is.
Example: $-6 < -1$

❸ Subtracting a negative number is like adding.
Example: $10 - (-4) = 14$

❹ Squaring a negative eliminates it, but cubing does not.
Example: $(-3)^2 = 9$, but $(-3)^3 = -27$

❺ Skill 37 review: Anything times zero equals zero.
Example: $(-16)(3)(0) = 0$

That's it. Know these and you will gain points, impress friends, and in general be a more attractive person.

Let's use this stuff on the question from the Pretest.

46. Suppose $0 < b < 1$. Which of the following has the greatest value?

   F. b^2 **G.** b^3 **H.** $\log b$ **J.** $|b|$ **K.** b^{-1}

Solution: This is a great question. Lots of good review, and it's a classic ACT question. The ACT loves to see if you pick up on this weird behavior stuff. Let's just "Make It Real" and see what happens. Let's say $b = 0.5$, and use your calculator to try each answer choice, looking for the greatest value:

F. $b^2 = (0.5)^2 = 0.25$ **G.** $b^3 = (0.5)^3 = 0.125$ **H.** $\log b = \log 0.5 = -0.3$
J. $|b| = |0.5| = 0.5$ **K.** $b^{-1} = (0.5)^{-1} = 2$.

There it is.

Correct answer: K

Misbehaving Numbers: Weird Number Behavior Drills

Easy

1 If $q = 2kp$ and $k = 0$, what is the value of q?

(A) -1

(B) 0

(C) 1

(D) 2

(E) 3

2 Which of the following numbers would be to the left of the number -1 on a number line?

(F) 8

(G) 4

(H) 0

(J) -0.5

(K) -4

Medium

3 If $x < -1$, which of the following best describes the relationship between x^2 and x^3?

(A) $x^2 > x^3$

(B) $x^2 < x^3$

(C) $x^2 = x^3$

(D) $x^2 = -x^3$

(E) $x^2 = |x^3|$

Hard

4 Suppose $-1 < b < 0$. Which of the following has the least value?

(F) b^2

(G) b^3

(H) $\tan b$

(J) $|b|$

(K) b^{-1}

5 If $-2 < x < -1$, then which of the following orders x, x^2, and x^3 from least to greatest?

(A) $x < x^2 < x^3$

(B) $x^2 < x^3 < x$

(C) $x^3 < x < x^2$

(D) $x < x^3 < x^2$

(E) $x^3 < x^2 < x$

Some kids are intimidated by the word "logarithm." But if you lived three thousand years ago, and you heard the word "logarithm," you'd just hear "ratio of numbers" and be like, "cool." "Logarithm" or "log" is just a fancy word for "ratio of numbers."

The ACT tests two kinds of log questions. The first are straight-up log questions that just ask the basics of what a log is. A log is just a fancy way of writing exponents. For example, $\log_5 25 = 2$ means that the little tiny 5 written next to the word "log" when taken to the power of 2 equals 25. $5^2 = 25$. You just need to memorize what each spot means. No different than memorizing that the m in $y = mx + b$ means slope.

The second log topic that the ACT tests is log expansion. Log expansion is easy. If you never even learned this before, you will right now, no problem. When you add logs, you multiply the numbers; and when you subtract logs, you divide the numbers. This makes sense since that's what we do for exponents, and logs are really just a form of exponents.

So, $\log_3 20 + \log_3 4 = \log_3(20 \cdot 4) = \log_3 80$, and $\log_3 20 - \log_3 4 = \log_3\left(\frac{20}{4}\right) = \log_3 5$.

Also exponents of logs can be moved in front of the word "log": $\log_4 3^2 = 2\log_4 3$. That's what you need to know.

Let's look at the question from the Pretest.

47. What is the real value of x in the equation $\log_2 16 = \log_4 x$?

    A. 2   **B.** 32   **C.** 64   **D.** 128   **E.** 256

Solution: Nice exponents review! $\log_2 16$ means 2 to the what equals 16. 2 to the 4 equals 16 ($2^4 = 16$), so we can substitute 4 for $\log_2 16$. We now have $4 = \log_4 x$, which means $4^4 = x$, and $x = 4 \cdot 4 \cdot 4 \cdot 4 = 256$. Whew!

Correct answer: E

ACT Math Mantra #47
A log is just a fancy way of writing exponents. For example,
$\log_5 25 = 2$ **means** $5^2 = 25$**.**

Log Drills

Medium

1 Which of the following is a value that satisfies $\log_n 144 = 2$?

(A) 72

(B) 36

(C) 24

(D) 18

(E) 12

2 Which of the following is a value that satisfies $\log_2 x = 5$?

(F) 32

(G) 16

(H) 10

(J) 7

(K) 3

Hard

3 Which of the following expressions is greatest?

(A) $\log_5 25$

(B) $\log_{25} 25$

(C) $\log_{25} 5$

(D) $\log_{625} 25$

(E) $\log_{125} 5$

4 If $\log_m n = a$ and $\log_m p = b$, then $\log_m (np)^3 = $?

(F) ab

(G) $3ab$

(H) $3a + b$

(J) $6a + b$

(K) $3a + 3b$

5 What is the real value of x in the equation $\log_3 x + \log_3 (9x) = 4$?

(A) 3

(B) 9

(C) 18

(D) 27

(E) 36

5 What is the real value of x in the equation $\log_4 16 - \log_6 36 = \log_8 x$?

(F) 1

(G) 2

(H) 10

(J) 20

(K) 32

Skill
48

Not So Complex Numbers

Math books call this topic *complex numbers*. Right off the bat, that's an unfortunate name. Could it be more intimidating? Who thought this name up? But that name is just fronting; complex numbers are really not very complex at all.

They are numbers that have a regular part, like 5 or –13, and an imaginary part, called *i*. What is *i*? $i = \sqrt{-1}$; that's why it's called imaginary, since $\sqrt{-1}$ does not exist. (Remember from Skill 12 that $\sqrt{-1}$ is undefined and the math police are on their way.) Like we didn't have enough numbers already. They said, "Let's imagine up some that don't even exist." Well, at least there is only one imaginary number, *i*.

So complex numbers have a normal part and an *i* part, like $2 + 2i$. These questions, like so many, are easy once you just know what *i* is. The key to these questions is that $i = \sqrt{-1}$ and therefore, $i^2 = -1$. **So, just treat *i* like a normal variable, and in the final step of a question, replace i^2 with –1.** That's it.

Let's look at the question from the Pretest.

48. What is $(i - 2)(i - 3)$?

   F. $5 - 5i$ **G.** $5 - 4i$ **H.** $5 + i$ **J.** 5   **K.** –1

Solution: No problem, and great FOIL review. Just FOIL $(i - 2)(i - 3)$ to get $i^2 - 3i - 2i + 6$ and collect like terms to get $i^2 - 5i + 6$. With normal FOILing you'd be done, but here there is one last step, the key to complex number questions. Since i^2 actually equals –1, we substitute –1 for i^2 to get $-1 - 5i + 6$ and collect like terms to get a final answer of $5 - 5i$.

Correct answer: F

ACT Math Mantra #48
The key to complex number questions is to treat *i* like a normal variable, and then in the final step, replace i^2 with –1.

Not So Complex Numbers Drills

Medium

① What is the product of $-2i$ and $(3i + 2)$?

Ⓐ -6

Ⓑ -5

Ⓒ $6 - 4i$

Ⓓ $6 + 4i$

Ⓔ $-5 - 4i$

② What is the square of the complex number $(i - 2)$?

Ⓕ -4

Ⓖ -3

Ⓗ $3 - 4i$

Ⓙ $4 - 4i$

Ⓚ $-4 - 4i$

Hard

③ In the complex numbers, $i^4 = ?$

Ⓐ $\sqrt{-1}$

Ⓑ -1

Ⓒ 0

Ⓓ 1

Ⓔ 2

④ In the complex numbers, $\dfrac{1}{1-i} \cdot \dfrac{1-i}{1+i} = ?$

Ⓕ $1 - i$

Ⓖ $1 + i$

Ⓗ $i - 1$

Ⓙ $\dfrac{2}{1-i}$

Ⓚ $\dfrac{1-i}{2}$

⑤ In the complex numbers, $\dfrac{5i}{1-i} \cdot \dfrac{1-i^2}{1+i} = ?$

Ⓐ $1 - i$

Ⓑ $1 + i$

Ⓒ $-1 - i$

Ⓓ $5i$

Ⓔ 5

You've now learned the 48 Skills that you need for the ACT. The Math Mantras remind you what to do when, what that girl who got a 36 does automatically. In Skills 49 and 50, let's make sure you've memorized the ACT Math Mantras. Cut out the flash cards provided at the end of this book and drill them until you are ready to teach them. Then do that. Once you're sure you've got 'em, check off the box next to each mantra.

☐ **Skill 1.** When you see <u>variables</u> or <u>unknowns</u> in the question and <u>numbers</u> in the answer choices, "Use the Answers." Convert fractions, π, or $\sqrt{}$ into decimals.

☐ **Skill 2.** When you see "... then $x = ?$" either complete the algebra or "Use the Answers."

☐ **Skill 3.** "What is y in terms of x and z" is just a fancy way of saying "Solve for y" or "Use algebra to get y alone."

☐ **Skill 4.** When you see the word "mean" or "average" on the ACT, use

$$\text{Average} = \frac{\text{sum}}{\text{number of items}}.$$

☐ **Skill 5.** When you see vertical angles, a linear pair, or a triangle, calculate the measures of all angles.

☐ **Skill 6.** When you see two parallel lines that are crossed by another line, eight angles are formed, and all of the bigger-looking angles are equal, and all of the smaller-looking angles are equal.

☐ **Skill 7.** When you see a triangle with two equal sides, mark the two opposite angles as equal; and when all sides of a triangle are equal, mark all angles 60°.

☐ **Skill 8.** When you see an expression like $(2x - 5)(5x - 4)$, multiply the first number and letter into the second parenthesis, multiply the second number and letter into the second parenthesis, and then collect matching terms.

☐ **Skill 9.** Anytime you see a math vocab term, underline it.

☐ **Skill 10.** Don't be intimidated by fancy vocabulary terms.

☐ **Skill 11.** To find the "least common multiple" or "lowest common denominator," "Use the Answers!"

☐ **Skill 12.** Don't be intimidated by fancy graphing terms.

☐ **Skill 13.** To find the slope or "rate of change" of a line, use Slope $= \frac{y_1 - y_2}{x_1 - x_2}$.

☐ **Skill 14.** An equation of a line in the form $y = mx + b$ is called the slope-intercept form, where m represents the slope of the line. Parallel lines have equal slopes, like $\frac{2}{3}$ and $\frac{2}{3}$. Perpendicular lines have negative reciprocal slopes, like $\frac{2}{3}$ and $-\frac{3}{2}$.

☐ **Skill 15.** The key to charts and graphs is to read the intro material and the "note" if there is one, and to expect an average, percent, and/or probability question about the data.

☐ **Skill 16.** $f(3)$ means "plug 3 in for x."
$f(m)$ means "plug m in for x."
$f(g(m))$ means "plug $g(m)$ in for x."

☐ **Skill 17.** To find the midpoint of two points, use the formula:
$$\text{Midpoint} = \left(\frac{x+x}{2}, \frac{y+y}{2} \right)$$
To find the distance between two points, use the formula:
$$\text{Distance} = \sqrt{(x-x)^2 + (y-y)^2}$$

☐ **Skill 18.** When a question with variables is way too theoretical, just "Make It Real."

☐ **Skill 19.** Memorize basic perimeter, area, and volume formulas. To use them, plug in what you know, and solve for the variable.

☐ **Skill 20.** The area of a donut equals the area of the big guy minus the area of the donut hole.

☐ **Skill 21.** 4 boys to 5 girls could also be expressed 5 girls to 9 students; a ratio can be a reduced version of the real numbers; and when you see a proportion on the ACT, cross-multiply.

☐ **Skill 22.** Relax when you see a matrix question, and treat the matrix like a normal chart. To add matrices, add corresponding numbers; to multiply, remember that the result will have as many rows as the first and as many columns as the second matrix being multiplied; and for any other operation, just follow the instructions that the question provides.

Here's the question from the Pretest.

49. The table below shows the results of a survey in which 180 high school students voted for their favorite movie. Each student received one vote. According to the graph, how many more students favored *Superbad* than *Wedding Crashers*?

Movie	Votes	♥ = 20 votes
Superbad	♥ ♥ ♥ ♥ ♥	
The Dark Knight	♥ ◖	
Wedding Crashers	♥ ◖	
Godfather	♥	

   A. 3　　　　**B.** 3.5　　　　**C.** 35　　　　**D.** 70　　　　**E.** 79.5

Solution: The legend tells us that each ♥ represents 20 votes. Therefore, *Superbad* received 100 votes and *Wedding Crashers* received 30, and $100 - 30 = 70$. Remember that half a symbol represents 10 votes, not half a vote.

Correct answer: D.

How to Think Like a Math Genius I Drills

Name the skill(s) that you can use, and then solve each question.

Easy

❶ According to Grandma Sylvia's chicken soup recipe, 3 pounds of chicken is needed to make 12 bowls of healing chicken soup. At this rate, how many pounds of chicken are needed to make 1 bowl?

Ⓐ 0.10
Ⓑ 0.25
Ⓒ 0.30
Ⓓ 1
Ⓔ 3

❷ What is the value of the expression $(2m - n)^2$ when $m = 3$ and $n = -2$?

Ⓕ 64
Ⓖ 49
Ⓗ 50
Ⓙ 16
Ⓚ 9

❸ If $\frac{35}{21} = \frac{5}{a}$, what is the value of a ?

Ⓐ 70
Ⓑ 35
Ⓒ 21
Ⓓ 6
Ⓔ 3

❹ Which of the following lists all the positive factors of 10 ?

Ⓕ 1, 10
Ⓖ 2, 5
Ⓗ 2, 5, 10
Ⓙ 1, 2, 5, 10
Ⓚ 10, 20, 30, 40

❺ For all x, $(4x - 3)^2 = ?$

Ⓐ $4x + 6$
Ⓑ $8x^2 + 9$
Ⓒ $16x^2 + 9$
Ⓓ $16x^2 - 12x + 9$
Ⓔ $16x^2 - 24x + 9$

❻ If two sides of a triangle are equal, which of the following could be the measures of its angles?

Ⓕ 30, 30, 90
Ⓖ 35, 45, 100
Ⓗ 35, 35, 110
Ⓙ 30, 60, 90
Ⓚ 45, 45, 45

7 If the mean of the perimeters of two shapes is 20, what is the sum of the perimeters of the two shapes?

(A) 0

(B) 10

(C) 15

(D) 20

(E) 40

10 The sides of a triangle are in the ratio $2:3:4$. If the shortest side is 10 inches long, what is the perimeter of the triangle?

(F) 6

(G) 9

(H) 17

(J) 45

(K) 90

Medium

8 In the diagram below, with angles as marked and $y = 78$ and $z = 23$, what is the value of x?

(F) 79

(G) 62

(H) 47

(J) 34

(K) 25

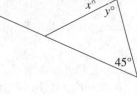

Hard

11 If a is a multiple of 3 and b is a multiple of 4, which of the following must be a multiple of 12?

(A) (ab) only

(B) $(3a + 4b)$ only

(C) $(4a + 3b)$ only

(D) (ab) and $(3a + 4b)$

(E) (ab) and $(4a + 3b)$

9 If $f(x) = 3x^3$ and $g(x) = x - 2$, what is the value of $f(-1) + g(5)$?

(A) −1

(B) 0

(C) 1

(D) 5

(E) 6

How to Think Like a Math Genius II

Learning mantras is like learning martial arts. Practice until they become part of you, until you follow them naturally: When you see a proportion, you cross-multiply; when you see a linear pair, you fill in the angles. . . . This will fundamentally change you as a math student. In fact, after ACT prep, many students begin to like math, they realize that they "get" it, and it stops being intimidating and becomes easy. I've even seen kids overcome serious math phobia with the mantras. Your ACT score and probably even your math class grades will go way up.

Here are the rest of the ACT math mantras. Check the box next to each one when you have mastered it.

☐ **Skill 23.** "Use the Diagram" to estimate an answer. When a diagram is not drawn to scale, redraw it. And when a picture is described but not shown, draw it! When estimating an answer, translate fractions, $\sqrt{}$, or π into decimals.

☐ **Skill 24.** When you see a right triangle, try $a^2 + b^2 = c^2$.

☐ **Skill 25.** When you see a 30°, 45°, or 60° angle in a right triangle, try using the special right triangles.

☐ **Skill 26.** Similar triangles have sides that are proportional.

☐ **Skill 27.** Translate word problems from English to math.

☐ **Skill 28.** Word problems are no-problems; translate English to math, and translate fractions to decimals.

☐ **Skill 29.** When something can be factored, FOILed, reduced, or simplified, do it. When you have two equations in a question, ask how they relate. And convert fractions to decimals.

☐ **Skill 30.** Memorize the laws of exponents.

☐ **Skill 32.** For the equation $y = mx + b$, the m is the slope and b is the y intercept.

☐ **Skill 33.** When you see an arrangement question, draw a blank for each position, fill in the # of possibilities to fill each position, and multiply. When an arrangement question mentions a "team of two," or specifically points out repeats, divide your result by 2.

☐ **Skill 34.** SohCahToa!

☐ **Skill 35.** When trig seems tough, "Use the Answers" or "Make It Real."

☐ **Skill 36.** When you see the word "probability," use the equation Probability $= \frac{\text{want}}{\text{total}}$.

☐ **Skill 37.** For questions like $(x + 4)(x - 3) = 0$, just "Use the Answers" or set each parenthesis equal to zero and solve for x.

☐ **Skill 38.** For the equation $y = ax^2 + bx + c$, the a tells whether the U shaped graph opens up or down, and the c is the y intercept. For the equation $y = (x - h)^2 + k$, the h and k give the coordinates of the vertex of the graph (h, k). The vertex is the highest or lowest point of the graph and is therefore also called the maximum or minimum point.

☐ **Skill 39.** The equation for a circle is $(x - h)^2 + (y - k)^2 = r^2$, where (h, k) is the center and r is the radius of the circle.

☐ **Skill 40.** If one side of a triangle is the diameter of a circle, and the opposite vertex is on the circle, then the triangle is right, with its right angle opposite the diameter.

☐ **Skill 41.** When you see absolute value on the ACT, "Use the Answers" or "Make It Real," and remember that absolute value means, "Ditch the negative sign."

☐ **Skill 42.** An arithmetic sequence is a sequence of numbers where a certain number is added to each term to arrive at the next, like 3, 7, 11, 15, 19; and a geometric sequence is a sequence of numbers where a certain number is multiplied by each term to arrive at the next, like 3, –6, 12, –24, 48.

☐ **Skill 43.** For a Fahrenheit/Celsius conversion question, when you are given degrees Celsius, just plug in and simplify; but when you are given degrees Fahrenheit, you can either do the algebra or "Use the Answers."

☐ **Skill 44.** "Careless errors are bad mmmkay," so underline all vocabulary words and remember to distribute the negative.

☐ **Skill 47.** A log is just a fancy way of writing exponents. For example, $\log_5 25 = 2$ means $5^2 = 25$.

☐ **Skill 48.** The key to complex number questions is to treat i like a normal variable, and then in the final step, replace i^2 with –1.

Here's the question from the Pretest.

50. Javier earned the following 7 test scores. What is the median?
85, 92, 82, 94, 90, 80, 79

   F. 80   **G.** 82   **H.** 85   **J.** 86   **K.** 90

Solution: This question would be ranked "medium" or even "hard," but if you just know what "median" means, it's totally easy! Median is the middle number in a list of numbers. To find the median, rewrite the list in order: 79, 80, 82, 85, 90, 92, 94. Then cross out a number on each side until you get to the one in the middle. (If there are two middle numbers, then the median is the average of the two.) Here the middle number is 85.

Correct answer: H

How to Think Like a Math Genius II Drills

Name the skill(s) that you can use, and then solve each question.

Easy

❶ Which of the following is equivalent to $2(3x^4)^3$?

   Ⓐ $6x^7$

   Ⓑ $27x^7$

   Ⓒ $27x^{12}$

   Ⓓ $54x^7$

   Ⓔ $54x^{12}$

❷ What is the length of the hypotenuse of a right triangle with legs that are 7 and 24 inches long, respectively?

   Ⓕ $\sqrt{17}$

   Ⓖ $2\sqrt{6}$

   Ⓗ 7

   Ⓙ 25

   Ⓚ 31

❸ Kayla purchased a new phone at 31% off its original price. If she paid $34 for the phone, which of the following is an equation that could be used to find x, the original price of the phone?

   Ⓐ $34 = 1.31x$

   Ⓑ $31 = x - 0.34x$

   Ⓒ $31 = x - 0.66x$

   Ⓓ $34 = x - 0.31x$

   Ⓔ $x = x - 0.31$

❹ What are the values for x that satisfy the equation $x^2 - 5x - 14 = 0$?

   Ⓕ -2 and 9

   Ⓖ -2 and -9

   Ⓗ 2

   Ⓙ 2 and -7

   Ⓚ -2 and 7

Medium

❺ In the standard (x, y) coordinate plane, the midpoint of $\overline{MN}$ is $(-3, 5)$. If point M is located at $(2, 3)$, what are the coordinates of point N?

   Ⓐ $(-8, 7)$

   Ⓑ $(7, 1)$

   Ⓒ $(-1, 8)$

   Ⓓ $(-6, 15)$

   Ⓔ $(0, 0)$

❻ For right triangle $\triangle ABC$, the sides measure 6, 8, and 10 inches. Which of the following equations could be used to solve for smallest angle, x, which is opposite the smallest side?

   Ⓕ $\sin x° = 0.6$

   Ⓖ $\sin x° = 0.8$

   Ⓗ $\tan x° = 0.8$

   Ⓙ $\sin x° = 0.75$

   Ⓚ $\tan x° = 0.6$

7 If $m < 0$, which of the following is greatest?

Ⓐ m

Ⓑ $3m$

Ⓒ $6m$

Ⓓ $9m$

Ⓔ It cannot be determined from the information given.

8 If triangle MNO is similar to triangle PQR shown below, which of the following could be the measures of the three sides of triangle MNO?

Ⓕ $3\sqrt{3}, 6\sqrt{3}, 12$

Ⓖ $6, 6\sqrt{3}, 12$

Ⓗ $6, 6, 12$

Ⓙ $3\sqrt{3}, 6\sqrt{3}, 6$

Ⓚ $12, 12, 6$

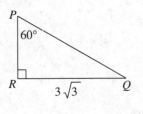

9 If 4 times a number p is subtracted from 20, the result is negative. Which of the following gives the possible value(s) for p?

Ⓐ 0 only

Ⓑ 5 only

Ⓒ 20 only

Ⓓ All $p < 5$

Ⓔ All $p > 5$

10 Which of the following could be the equation of the graph below?

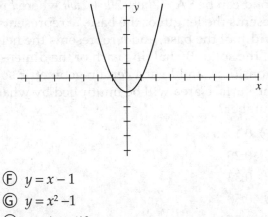

Ⓕ $y = x - 1$

Ⓖ $y = x^2 - 1$

Ⓗ $y = (x - 1)^2$

Ⓙ $y = (x + 1)^2$

Ⓚ $y = x^2$

11 The inequality $3(3x - 2) < 6 - (5x - 2)$ is equivalent to

Ⓐ $x < 7$

Ⓑ $x < 5$

Ⓒ $x < 3$

Ⓓ $x < 1$

Ⓔ $x < -1$

Hard

12 A formula for the surface area of a rectangular solid can be SA = 2*lw* + 2*lh* + 2*wh* where *l* represents the length of the base, *w* represents the width of the base, and *h* represents the height of the solid. By halving each of the dimensions (*l*, *w*, and *h*) of a certain geometric solid, the surface area will be multiplied by what number?

F 0.1

G 0.25

H 0.33

J 0.50

K 0.75

13 The dartboard shown below has a small circle, with radius 3.5, inside a larger circle, with radius 5. When Matty randomly throws a dart at the board, which of the following is closest to the probability that it will land in the shaded region?

A $\frac{1}{12}$

B $\frac{1}{6}$

C $\frac{1}{2}$

D $\frac{3}{4}$

E $\frac{5}{6}$

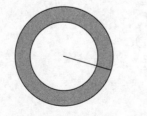

14 If the width of a little league baseball must satisfy the inequality $|w - 3| \leq 0.12$, what is the widest that a ball can be?

F 2.12

G 2.88

H 3.00

J 3.12

K 3.15

Brian's Friday Night Spiel
Recommendations for the Days Preceding the Test

Studies show that sleeping and eating healthfully two days before the test (or any important event) is as important as sleeping and eating healthfully the night before. So Thursday night eat a healthy dinner and go to bed early—not so early that you're lying in bed at 7:00 p.m. tense, hungry, and staring at the cracks in the ceiling, but normal early, maybe 10:00 p.m.

Friday, have a normal day, no need to cram or stress. If you have completed this book and one or more timed practice tests, you are ready. Go to school, play sports or whatever you do after school, have a healthy dinner, and do something fun and relaxing. Don't hang out with anyone who stresses you out or obsesses over the test. Read, play a game, or watch a funny movie—I recommend *Fletch*, *Wedding Crashers*, or *40 Year-old Virgin*—and go to bed at a sensible time. If you live in a household where, in the morning, everyone roams the house screaming for a clean shirt and their car keys, then gather your snack, drink, admission ticket, ID, pencils, watch, and calculator in the evening.

You should eat breakfast and pack a snack because it's a long day and you have to feed the brain. I recommend a cheese sandwich or two Luna bars; they are high in protein and not too high in sugar, good brain food. If you need an extra special boost, in India some people take a few drops of almond oil with breakfast on the morning of a test.

Let's see if you got all that:

Two nights before test day, you should

A. skip dinner and go to bed at 6 p.m.
B. ask a friend to motivate you by shouting, "You are nothing!"
C. repeat the mantra, "Your future depends on this test"
D. eat two whole live lobsters for omega-3 fatty acids
E. have a healthy dinner, relax, and sleep well

Solution: Relax and sleep well, you are prepared. Now, go get'em!!

Correct answer: E

Test Day Checklist:	
2 Protein bars	#2 pencils
Beverage	Calculator
Your admission ticket	Watch (to keep track of time)
Photo ID (driver's license, school ID, or passport. See www.act.org for more options.)	

Brian's Friday Night Spiel Drills

Here is your last drill section. Your last assignment is to be able to stay relaxed, even under pressure. So here is a little tool that you can use anytime, even during the test.

In the 1970s Herbert Benson, a researcher at Harvard Medical School, published work on what he called the *relaxation response*, a physiological response where the body and mind relax. Benson reported that the relaxation response was triggered by practicing 20 minutes of a concentration exercise, basically meditation. Apparently, Yale, always in competition with Harvard, decided to one-up them. "We need a way to trigger the relaxation response but in less than Benson's 20 minutes!" they might have bemoaned. They researched, and they tried as hard as they could to relax; it was quite stressful. Finally, someone came up with the following goofy exercise. And it is goofy, but the thing is, it works! It totally works. Do it and you'll see.

Follow these steps:

1. Breathing through your nose, become aware of your breath.
2. Relax your shoulders and face.
3. Allow your exhale to be longer than your inhale.
4. Now, drop your shoulders and head and smile, and then bring your head back up.
5. Repeat: drop your shoulders and head and smile, and then bring your head back up.
5. Notice how you feel.

That's it! Anytime you feel stressed, even during the test, try this very simple exercise to trigger your "relaxation response."

Score is tied: Yale, 1. Harvard, 1.

> Sucking all the marrow out of life doesn't mean choking on the bone.
>
> John Keating, *Dead Poet's Society* (Touchstone Pictures, 1989)

Want to break 30? Here are the six steps to do it.

But first you have to promise me that you are doing this because you want to, and not out of some obsessive, sleep-doesn't-matter, gotta-please-my-parents, if-I-don't-go-to-Tufts I'm-nothing misunderstanding. Strive to do well, yes. Also stay balanced. Sleep. Eat well. Exercise. Be true to yourself. Be brave. Be honest. Be relaxed. Breathe. And from that place, give it all that you got.

1 Make sure you understand all 50 Skills. Don't just look at them and say, "Yeah, I can do that." Practice. Do the drills. Make sure you can get every question correct on every skill. If you can't, reread the section, read the solutions, and keep redoing the drills until they make perfect sense. Then teach them to a friend.

2 Memorize the careless errors to avoid. Keep them in mind as you drill, until avoiding careless errors becomes second nature.

3 Master Posttests I and II (Posttest II is online at www.MH-ACT-TOP50Math.com). Take each test, read the solutions, and redo any questions that you missed.

4 Once you've mastered Posttests I and II, take Posttest III (online), which consists of 50 very hard questions. Take the test, read the solutions, and redo any questions that you missed. When you master these 50 very challenging questions, you are ready to break 30!

5 Then take **at least** one timed ACT (see "Now What?" page 127). Score and review the test. Keep taking timed practice tests until you confidently and consistently break 30.

5 Finally, remember the words of quantum physics and of Jedi Master Yoda, "Do, or do not. There is no try." (*Star Wars Episode V: The Empire Strikes Back*, 20th Century Fox, 1980)

Easy, Medium, Hard, and Guessing Revisited

Let's revisit what I told you way back at the beginning of the book. It will probably make even more sense now.

The ACT is not graded like a math test at school. If you got only half the questions right on an algebra midterm, that'd be a big fat F. On the math ACT, half the questions right is a 21, the average score for kids across the country. If you got 67% of the questions right, that'd be a D+ in school, but a nice 25 on the ACT, the average score for admission to great schools like Goucher and University of Vermont. And 87% correct, which is a B+ in school, is a beautiful 30 on the ACT, and about the average for kids who got into Tufts, U.C. Berkeley, University of Michigan, and Emory.

The math questions on the ACT are organized in order of difficulty, from easiest to hardest. You do NOT lose points for wrong answers; there is no penalty for guessing. So, of course, you must put an answer for every question, even ones that you do not get to. In fact, you only need to get to the very hardest questions if you are shooting for 31+.

So if you need half correct, or 67% correct, don't rush through the easiest just to get to the hardest ones. For example, if you only make it to number 50 out of 60, you can guess on the last ten. In school you might need to finish tests in order to do well, here you do not. **You only need to get to the very hardest questions if you are shooting for 31+.**

Remember, on the ACT you lose no points for wrong answers. Even if you are running out of time at question #40 out of 60, you must budget a few minutes to fill in an answer for the last 20 questions. It'd be crazy not to. Statistically, if you randomly fill in the last 20 ovals, you'll get 4 correct. That's worth about 2 points (out of 36) on your score! So keep an eye on the clock, and when there are a few minutes left, choose an answer for each remaining question. Of course, when you have completed this book, you'll probably be able to finish the test and rarely even need to guess! When you do feel stumped, take another look and ask yourself, "Which Math Mantra can I use?"

Now What?

Take the Posttest. Check your answers and review the skills for any questions that were difficult. Then take Posttest II, found online at www.MH-ACT-TOP50Math.com. If you are shooting for 31+, take Posttest III (also available online), which contains 50 "hard" questions.

After you have completed the posttests, go to your guidance office and pick up the free packet entitled "The ACT: Preparing for the ACT," which contains a full practice test with answers and scoring instructions. Or you can download a free test at www.actstudent.org.

Take the test as a dress rehearsal; get up early on a Saturday, time it, use the answer sheets, and fill in the ovals. If you have competed this book, you will find that you are very well prepared. Correct and score the test, and review whatever you got wrong. Figure out which ACT Math Mantras you could have used to get them right.

If you have some time, purchase *The Real ACT Prep Guide*. It contains five practice tests. Take one practice test per week as a dress rehearsal. Take it when you are relaxed and focused. We want only your best work. Less than that will earn you a lower score than you are capable of and is bad for morale. Score each test and review whatever you got wrong. Figure out which ACT Math Mantras you could have used to get them right.

Now, you are ready, you beautiful ACT monster. Go get 'em!

Posttest I

This posttest contains 50 questions that review our 50 Skills. Take the test. Then check your answers and review the skills for any questions that were difficult.

1 A sack of sweet yams could be equally divided among 4, 5, or 6 families. What is the smallest number of yams that could be in the sack?

Ⓐ 24
Ⓑ 30
Ⓒ 56
Ⓓ 116
Ⓔ 120

2 If $4^{(x+1)} = 8^x$, then $x = ?$

Ⓕ 1
Ⓖ 2
Ⓗ 4
Ⓙ 8
Ⓚ −7

3 If $D = RT$, then which of the following is an expression for T in terms of D and R?

Ⓐ DR

Ⓑ $D - R$

Ⓒ $\dfrac{R}{D}$

Ⓓ $\dfrac{D}{R}$

Ⓔ $R - D$

4 Sora earned $5 on Monday. On each following day, through Friday, she earned 1 more dollar than the preceding day. What was her mean earnings for the five-day period, Monday through Friday?

Ⓕ 4
Ⓖ 5
Ⓗ 6
Ⓙ 7
Ⓚ 8

5 The measure of ∠BFE in the figure below is 138°. The measures of 3 angles are given in terms of m, in degrees. What is the measure of ∠BFD?

Ⓐ 110
Ⓑ 115
Ⓒ 120
Ⓓ 125
Ⓔ 130

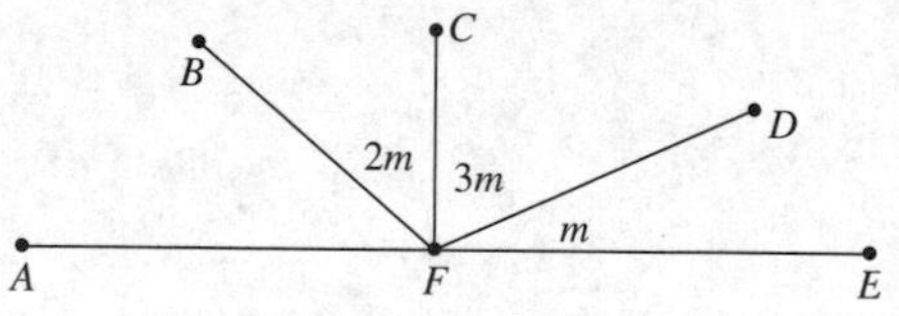

6 In the figure below, if $a \parallel b$ and the measure of $x = 25$, what is the measure of y?

Ⓕ 25
Ⓖ 50
Ⓗ 90
Ⓙ 108
Ⓚ 133

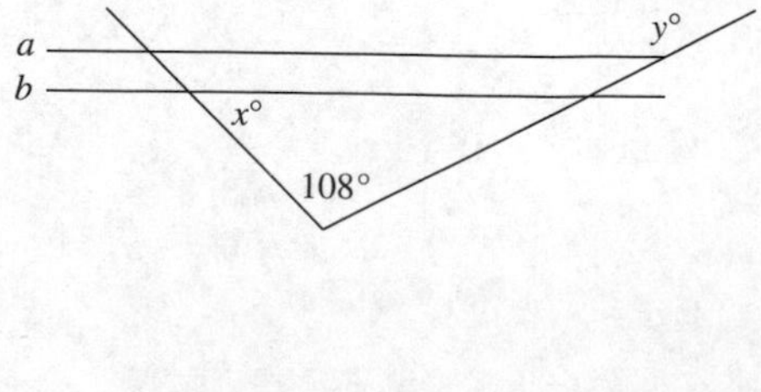

7 In $\triangle QRX$, $\overline{RX} = \overline{XQ}$ and the measure of $\angle Q$ is 34°. What is the measure of $\angle X$?

  Ⓐ 112°

  Ⓑ 56°

  Ⓒ 54°

  Ⓓ 34°

  Ⓔ 17°

8 For all x, $(5x - 2)^2 = ?$

  Ⓕ $5x^2 + 4$

  Ⓖ $25x^2 - 2$

  Ⓗ $25x^2 - 10x + 2$

  Ⓙ $25x^2 - 20x + 4$

  Ⓚ $x^2 - 25x - 2$

9 Which of the following lists gives the 3 largest prime numbers that are less than 50 ?

  Ⓐ 5, 7, and 11

  Ⓑ 7, 11, and 13

  Ⓒ 41, 43, and 47

  Ⓓ 39, 43, and 47

  Ⓔ 43, 47, and 49

10 Which of the following lists the factors of 36 ?

  Ⓕ 1, 36

  Ⓖ 2, 3

  Ⓗ 36, 72, 144

  Ⓙ 1, 2, 3, 4, 6, 9, 12, 18, 36

  Ⓚ The empty set

11 What is the lowest common denominator of $\frac{1}{3}$, $\frac{1}{8}$, and $\frac{1}{12}$?

  Ⓐ 3

  Ⓑ 24

  Ⓒ 36

  Ⓓ 96

  Ⓔ 288

12 In the xy coordinate plane below, which of the following points has coordinates (x, y) such that $x = y - 2$?

  Ⓕ A

  Ⓖ B

  Ⓗ C

  Ⓙ D

  Ⓚ E

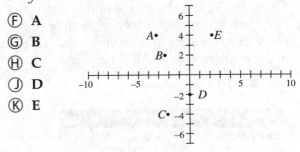

13 If the line through the points $(5, -3)$ and $(-2, p)$ is parallel to the line $y = -2x - 3$, what is the value of p?

(A) 11
(B) 4
(C) 0
(D) -10
(E) -17

(A) 0.30
(B) 0.25
(C) 0.20
(D) 0.15
(E) 0.10

14 In the standard (x, y) coordinate plane, what is the slope of the line parallel to the line with equation $6x + 8y = -12$?

(F) -6
(G) $-\frac{4}{3}$
(H) $-\frac{3}{4}$
(J) -2
(K) 4

16 The number of yoga postures that Ben knows after a series of yoga classes can be modeled by the function $P(c) = 3c + 8$, where c represents the number of classes that he has attended. Using this model, how many postures would you expect him to know after 10 classes?

(F) 3
(G) 8
(H) 10
(J) 30
(K) 38

15 The graph below shows the number of Texas Hold 'em tournaments won by each of four players. According to the graph, what fraction of the tournaments did Reisner win?

Key: ♣ = 10 tournaments

Player	Games Won
Molinaroli	♣ ♣ ♣
Merlin	♣ ♣ ♣
Lang	♣ ♪
Reisner	♣ ♣ ♪

17 What is the distance, in coordinate units, between the points $(-4, 3)$ and $(7, -2)$ in the standard (x, y) coordinate plane?

(A) $\sqrt{14}$
(B) $\sqrt{98}$
(C) $\sqrt{146}$
(D) 15
(E) 21

18 In the equation $y = \dfrac{5}{2 + x}$, x represents a positive integer. As x gets larger and larger without bound, the value of y

(F) approaches 5

(G) approaches 2

(H) approaches 1

(J) approaches 0

(K) remains unchanged

19 If a completely fenced in square lot requires 80 feet of fence, what is the area, in square feet, of the lot?

(A) 6400

(B) 1600

(C) 400

(D) 200

(E) 20

20 In the figure below, what is the area of the shaded region?

(F) 17

(G) 32

(H) 39

(J) 49

(K) 54

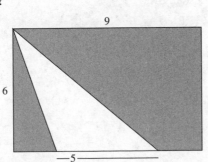

21 If the numerator of a fraction is 4 more than the denominator and the fraction equals $\dfrac{5}{3}$, what is the numerator of the fraction?

(A) 3

(B) 6

(C) 9

(D) 10

(E) 14

22 $\begin{bmatrix} p & q \\ r & s \end{bmatrix} - 2\begin{bmatrix} w & x \\ y & z \end{bmatrix} = ?$

(F) $\begin{bmatrix} 2pw & 2qx \\ 2ry & 2sz \end{bmatrix}$

(G) $\begin{bmatrix} 2 & 2 \\ 2 & 2 \end{bmatrix}$

(H) $\begin{bmatrix} p + 2w & q + 2x \\ r + 2y & s + 2z \end{bmatrix}$

(J) $\begin{bmatrix} p - 2w & q - 2x \\ r - 2y & s - 2z \end{bmatrix}$

(K) undefined

23 In a standard (x, y) coordinate plane, three points Q, R, and S have coordinates $(5, 2)$, $(-1, 5)$, and $(5, b)$. If the three points determine a right triangle, what is the value of b ?

(A) 0

(B) 2

(C) 4

(D) 5

(E) 6

24. The lengths of the sides of a right triangle are consecutive integers, and the length of the shortest side is x. Which of the following could be used to solve for x?

F. $(x)^2 + (x+1)^2 = (x+2)^2$

G. $x + x - 3 = x$

H. $(x)(x+1) = (x+2)^2$

J. $(x+2) - (x+1) = x$

K. $(x)^2 + (x+2)^2 = (x+4)^2$

25. Which of the following sets of 3 numbers could be the side lengths, in meters, of a 45°–45°–90° triangle?

A. $2, 2, 2$

B. $2, 2, 2\sqrt{2}$

C. $2, 2\sqrt{2}, 2\sqrt{2}$

D. $2, 2\sqrt{2}, 2\sqrt{3}$

E. $2, 2\sqrt{3}, 4$

26. Edward casts a shadow 3 meters long. At the same time, Bella casts a shadow 2.5 meters long. If Edward is 2 meters tall, which of the following is closest to Bella's height, in meters?

F. 0.6

G. 1.2

H. 1.7

J. 2.6

K. 3.4

27. 71 is what percent of 1420?

A. 2%

B. 5%

C. 20%

D. 50%

E. 700%

28. What is the total cost of 4.3 pounds of flour at $1.10 per pound and 0.5 pounds of baking soda at $0.82 per pound?

F. $5.14

G. $4.27

H. $3.34

J. $2.46

K. $1.78

29. If $x^2 - y^2 = 72$ and $x - y = 8$, what is the value of $x + y$?

A. 8

B. 9

C. 36

D. 72

E. 80

30 What is the product of $3m^4$ and $2b^4$?

   (F) $5mb^8$

   (G) $6mb^8$

   (H) $6(mb)^4$

   (J) $5(mb)^8$

   (K) $6(mb)^8$

31 What is the sum of $4x^2$ and $5x^2$?

   (A) $9x^4$

   (B) $9x^2$

   (C) $4x^2 + 5x^2$

   (D) $20x^4$

   (E) $20x^2$

32 If Brian spends 7 minutes warming up and 3 minutes per yoga posture, which of the following equations expresses the number of minutes Brian spends on warm-ups and postures when he does p postures?

   (F) $y = 7$

   (G) $y = 7p + 7$

   (H) $y = 10p$

   (J) $y = 3p + 7$

   (K) $y = 4p + 4$

33 On the first day of Professor Gil's *Reading Shakespeare* class, six students will select a seat from the six seats at a table in the classroom. How many different sitting arrangements of the six students are possible?

   (A) 6

   (B) 72

   (C) 144

   (D) 720

   (E) 1440

34 In the right triangle shown below, which of the following statements is true about $\angle L$?

   (F) $\cos L = \dfrac{8}{17}$

   (G) $\sin L = \dfrac{8}{17}$

   (H) $\tan L = \dfrac{8}{17}$

   (J) $\cos L = \dfrac{15}{8}$

   (K) $\sin L = \dfrac{15}{17}$

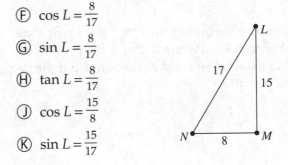

35 Which of the following expressions could be used to find the length of the side marked with the question mark in the triangle shown below, with measurements as marked?

(Note: The law of cosines states that for any triangle with vertices A, B, and C and the sides opposite those vertices with lengths a, b, and c, respectively, $c^2 = a^2 + b^2 - 2ab \cos C$.)

(A) $41^2 + 32^2 - 2(41)(32) \cos 67$

(B) $35^2 + 78^2 - 2(41)(32) \cos 67$

(C) $35^2 + 78^2 - 67(41)(32) \cos 2$

(D) $41^2 + 32^2 - 67(35)(67) \cos 41$

(E) $35^2 + 78^2 - 2(41)(32) \cos 32$

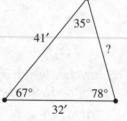

36 There are 16 buttons in a bag: 8 are red, 4 are blue, and 4 are white. What is the probability that a button selected at random from the bag is NOT white?

(F) $\frac{1}{6}$

(G) $\frac{1}{3}$

(H) $\frac{2}{3}$

(J) $\frac{3}{4}$

(K) 1

37 What is the product of the 2 solutions of the equation $x^2 + 2x - 24 = 0$?

(A) −24

(B) −10

(C) −2

(D) 8

(E) 14

38 What are the coordinates for the minimum point in the graph of $y - 5 = (x - 4)^2$?

(F) (4, 5)

(G) (−4, 5)

(H) (4, −5)

(J) (−4, −5)

(K) (0, 0)

39 A circle in the standard (x, y) coordinate plane has center (4, 9) and radius of 9 coordinate units. Which of the following is an equation of the circle?

(A) $(x - 4)^2 - (y - 3)^2 = 9$

(B) $(x + 4)^2 + (y + 9)^2 = 9$

(C) $(x - 4)^2 - (y - 9)^2 = 81$

(D) $(x - 4)^2 + (y - 9)^2 = 81$

(E) $(x + 4)^2 - (y - 9)^2 = 81$

40 The ratio of the radii of two circles is $2:3$. What is the ratio of their areas?

(F) $2:3$

(G) $4:9$

(H) $8:27$

(J) $2:\pi$

(K) $\pi:3$

41 Which of the following are solutions to $|q-5|=10$?

(A) 15 only

(B) -5 only

(C) -15 only

(D) -5 and -15 only

(E) 15 and -5 only

42 What is the next term after 12 in the geometric sequence $-\dfrac{3}{2}, 3, -6, 12, \ldots$?

(F) $\dfrac{1}{2}$

(G) 24

(H) 0

(J) -24

(K) -48

43 137°F is the point at which ethanol will boil. Since Fahrenheit and Celsius temperatures are related by the formula $F=\dfrac{9}{5}C+32$, to the nearest degree, which of the following is the boiling point of ethanol in degrees Celsius?
(Note: $F=\dfrac{9}{5}C+32$)

(A) 58°C

(B) 76°C

(C) 105°C

(D) 247°C

(E) 279°C

44 $(5x+2)-(8x-5)$ is equivalent to

(F) $(-3x-3)$

(G) $(-3x+7)$

(H) $(3x+3)$

(J) $(3x-7)$

(K) $(x-3)$

45 When $x=-3$, what is the value of the equation $y=2x^2+4$?

(A) -2

(B) -14

(C) 12

(D) 22

(E) 40

46 If $(x - 2)(x + 3) = 0$, then x could equal

   I. -3

   II. 0

   III. 2

   Ⓕ II only

   Ⓖ I and II

   Ⓗ I and III

   Ⓙ II and III

   Ⓚ I, II, and III

47 Which of the following is a value that satisfies $\log_6 216 = x$?

   Ⓐ 0

   Ⓑ 1

   Ⓒ 2

   Ⓓ 3

   Ⓔ 4

48 If the product of 3 and $(2i^2 + x)$ equals -9, then $x = ?$

   Ⓕ 0

   Ⓖ -1

   Ⓗ -2

   Ⓙ i

   Ⓚ $2i$

49 What is the slope of a line parallel to the line through $(-6, 3)$ and $(5, 8)$ in the standard (x, y) coordinate plane?

   Ⓐ -2

   Ⓑ $\frac{11}{5}$

   Ⓒ $\frac{5}{11}$

   Ⓓ 5

   Ⓔ 16

50 If $m = -2$ and $n > 0$, which of the following is greatest?

   Ⓕ mn

   Ⓖ $m^2 n$

   Ⓗ $m^3 n$

   Ⓙ $m^4 n$

   Ⓚ $m^5 n$

Solutions

Pretest

Solutions are on each skill page.

Skill 1 (page 13)

Use the Answers

1. **A** You can answer this question quickly and easily by using simple algebra and solving for x. If, however, that is difficult or daunting, "Use the Answers." Just plug each choice in to see which one makes the equation work. Choice A is correct: $3(-2) - 4 = -10$.

2. **F** This question looks pretty tough to some students, but "Use the Answers" makes it EASY! Just use your calculator to convert each $\sqrt{}$ into a decimal, and then see which answer choice works. (We will also briefly discuss the "math class" way to answer this question in Skill 29.)

$$\sqrt{40} + \sqrt{160} = ?$$
$$6.32 + 12.65 = ?$$
$$18.97 = ?$$

Translate each answer choice. Whichever is closest to 18.97 is correct. Choice F is correct, since $6\sqrt{10} = 18.97$.

3. **D** You can answer this question by using algebra and solving for n. Or "Use the Answers" to make this "medium" question EASY! Just use your calculator to convert each fraction to a decimal, and then plug each answer choice into the question to see which one makes the equation work. Choice D is correct:

$$n - \frac{2}{3} = \frac{11}{27} \Rightarrow n - 0.6667 = 0.407 \Rightarrow$$
$$n = 1.07 = \frac{29}{27}$$

4. **H** You can answer this question by using algebra and solving for p, but "Use the Answers" makes this "medium" question EASY! Just plug in each choice to see which one makes the inequality work. Choice H is

correct since $\frac{15}{18} = 0.83333$ and $\frac{10}{12} = 0.83333$.

Notice that the $\geq$ sign allowed us to use 10 as an answer, since the answer could be "greater than or **equal to**." If there had only been a > sign, the answer would be 9.

5. **D** You can answer this question by setting up two equations and using algebra to solve for the variables, but "Use the Answers" makes this "hard" question EASY! Just plug in each choice to see which one makes the situation work. Choice D is correct since 42 correct questions earn 168 points, and 8 incorrect questions subtract 8 points, which equals 160 points.

Skill 2 (page 15)

Super Easy Algebra

1. **C** $6x + 6 = 36 + 3x$

$-3x$	$-3x$	subtract $3x$
$3x + 6 = 36$		
-6	-6	subtract 6
$3x$	$= 30$	
$\div 3$	$\div 3$	divide by 3
x	$= 10$	

2. **G** $3(x + 4) + 2x = 2(x - 2) - 4$

$3x + 12 + 2x = 2x - 4 - 4$		distribute
$5x + 12 \quad\quad = 2x - 8$		collect like terms
$-2x$	$-2x$	subtract $2x$
$3x + 12$	$= -8$	
-12	-12	subtract 12
$3x$	$= -20$	
$\div 3$	$\div 3$	divide by 3
x	$= -\dfrac{20}{3}$	

3. **C** This is still simple algebra, but hidden in a word problem. No problem! We just hunt for the info. Remember they do not try to trick you; if you hunt, it will be straightforward. The question indicates that we want to know what F equals when H equals 0.16, so we plug in 0.16 for H and then solve for F—it's just a normal algebra question. And, as always, you can also "Use the Answers" if you prefer!

$$0.16 = \frac{3}{5}F - 0.02$$

$$\underline{+0.02 \qquad + 0.02} \qquad \text{add } 0.02$$

$$0.18 = \frac{3}{5}F$$

$$\left(\frac{5}{3}\right)(0.18) = \left(\frac{5}{3}\right)\left(\frac{3}{5}\right)F \qquad \text{multiply both sides by } \frac{5}{3}$$

$$0.3 = F$$

4. **K** Just subtract x from both sides to get $-2 < -4$ which is not true, so the equation never works, and the answer is the empty set, which is fancy math vocab for "no numbers work."

Skill 3 (page 17)

Super Easy Algebra—Algebraic Manipulation

1. **C** R in terms of I, P and T" means solve for R. In other words, get R alone.

$$I = PRT \qquad \text{divide both sides by } PT$$
$$\frac{I}{PT} = R$$

2. **J** This is a simple "$P = ?$" question where our answer will have numbers and variables in it. Just solve for P:

$$N = 5P + 4$$
$$\underline{-4 \qquad -4} \qquad \text{subtract } 4$$
$$N - 4 = 5P$$
$$\underline{\div 5 \qquad \div 5} \qquad \text{divide both sides by } 5$$
$$\frac{N-4}{5} = P$$

3. **E** "y in terms of x" means solve for y, get it alone. But we want it in terms of x, so we need to solve the first equation for m and then plug that into the second equation. So solving $x = 3m - 4$ for m, we get $m = \frac{x+4}{3}$. And plugging that into the second equation, we have $y = 6 - \left(\frac{x+4}{3}\right) = \frac{18}{3} - \left(\frac{x+4}{3}\right) = \left(\frac{14-x}{3}\right)$.

Skill 4 (page 19)

"Mean" Means Average

1. **D** Add up the five prices and divide by 5 to find the average.

2. **J** Now that you know that "mean" is just a fancy word for "average," you're all set. Add up the three amounts and divide by 3 to find the average. Then choose the answer choice closest to your result.

3. **D** You can set up the average formula and solve for the third bowling score, or you can just "Use the Answers," trying each answer choice and seeing which one works. Try both methods and see which one feels more comfortable for you.

$$\text{Average} = \frac{\text{sum}}{\text{number of items}}$$
$$205 = \frac{180 + 210 + x}{3}$$

$$615 = 180 + 210 + x \qquad \text{multiply both sides by } 3$$
$$615 = 390 + x \qquad \text{add the numbers together}$$
$$225 = x \qquad \text{subtract } 390 \text{ from both sides}$$

4. **G** Identify the data that the question is referring to: "the average number of students enrolled per section in World History." The enrollments for the three sections of World History are 19, 25, and 19. So add these up and divide by 3 to get your answer.

5. **E** Simply draw blanks for the 9 numbers and place 42 in the middle blank. Then fill in the consecutive numbers above and below 42. "Consecutive" means numbers in a row, for example, 42, 43, 44, 45, 46. So the greatest number is 46.

Skill 5 (page 21)

The Six-Minute Abs of Geometry: Angles

1. **D** Honestly this question would usually come with a diagram that makes it easy. But I wanted to make sure that you would learn the term "supplement," which means the angle required to add up to 180°. So the supplement of a 42.5° angle is $180 - 42.5 = 137.5$. This question brings up another important issue. As soon as they see "Cannot be determined from the information" as a choice, many students immediately pick it as an answer. There are many urban legends surrounding this. Some say it is **always** the answer, and others say it is **never** the answer. **Truth**: It is **sometimes** the answer. Specifically, it is **sometimes** the answer on an "easy" or "medium" question and **rarely** the answer for a "hard" question.

2. **K** Always mark info from the question into the diagram, so mark $y = 45$ into the figure. And whenever you see a linear pair, solve the other angle, which in this case equals $180 - 45 = 135$. Now we have 4 of the 5 angles of a five-sided figure. Skill #5 tells us that a five-sided figure has 540°, so 540 minus the other angles equals the measure of angle z. $z = 540 - 135 - 133 - 139 - 64 = 69$.

3. **E** Label the info from the question into the diagram, $x = 60$. Whenever you see a vertical angle, mark the angle opposite to it. Now, we have two angles of the triangle, 50 and 60. To get the third, subtract these two angles from 180: $180 - 50 - 60 = 70$. Now, to get the value of y, which is a linear pair to 70, subtract 70 from 180: $180 - 70 = 110$. Careless Error Buster: Remember to finish the question. Don't get 70 and stop. Ask yourself, "Did I finish the question?"

4. **G** Label all info from the question into the diagram. Then fill in the linear pair to $\angle CDE$: $180 - 110 = 70$. Since $ABCD$ is a trapezoid, BC is parallel to AD and angle $\angle DCB = 110$ (Skill 6 Preview). Now, we have three of the angles of the trapezoid which, since it has four

sides, must add up to 360. So $360 - 110 - 49 - 70 = 131$. Finally, to get $\angle DBA$, subtract: $131 - 19 = 112$.

Skill 6 (page 23)

The Six-Minute Abs of Geometry: Parallel Lines

1. **D** In a pair of parallel lines there are only two kinds of angles, big and little, and the two add up to 180. Here y is a little angle and x is clearly a bigger angle, so they add up to 180, and $x = 180 - 37.5 = 142.5$.

2. **K** Parallel lines cut by a transversal make 2 kinds of angles, big and little. Clearly x is little and z is big. If $x = 73$, then $z = 180 - 73 = 107$.

3. **A** First, mark the info that is given in the question into the diagram. That always helps to make the question simpler and usually shows which strategies to use. Then use vertical angles, linear pairs, and triangles to calculate the measures of any other angles that you can. The vertical angle to z is 42.5, and the linear pair to x is 55, so now we have two angles of the triangle. $180 - 42.5 - 55 = 82.5$. And the linear pair of 82.5, which is y, is $180 - 82.5 = 97.5 = y$.

4. **G** Mark the info from the question into the diagram. Then mark all other angles that you can determine. Here x and y are a linear pair, so $x + y = 180$ and $x = 180 - 75 = 105$. Since z, x, and y are three different numbers, they cannot be the angles of two parallel lines, which are always at most two different numbers. Also, since x and z are both less than 90, lines p and q cannot be parallel—they are heading toward each other and will intersect. Only G is the true choice.

5. **C** Only choice C is **NOT** justifiable:

    A $\overline{YT}$ is perpendicular to $\overline{ZT}$—justifiable as marked

    B VZW is congruent to $\angle XTW$—justifiable, all small angles are equal in the pair of parallel lines

Ⓒ $\overline{XY}$ is congruent to $\overline{TX}$—NOT justifiable, we have no proof

Ⓓ $\triangle ZVW$ is similar to $\triangle TXW$—(Skill 26 Preview) yes, the angles of $\triangle ZVW$ are equal to the corresponding angles of $\triangle TXW$, and therefore the triangles are similar

Ⓔ $\overline{VZ}$ is parallel to $\overline{TX}$—justifiable, both are perpendicular to $\overline{VY}$ and therefore parallel to each other

Skill 7 (page 25)

The Six-Minute Abs of Geometry: Triangles

1. Ⓓ As soon as you see an equilateral triangle, mark each angle 60°. Then since x forms a linear pair with 60° angle $\angle NMO$, $x = 180 - 60 = 120$.

2. Ⓕ When you are given one angle in a linear pair, mark the other. So $\angle SPQ = 180 - 108 = 72$. And as soon as you see two sides of a triangle are equal, mark their opposite angles equal. So since $PS = SQ$, $\angle SPQ = \angle SQP$. We already determined that $\angle SPQ = 72$, so $\angle SQP$ must also equal 72. Then since the angles in a triangle must add up to 180, $\angle PSQ = 180 - 72 - 72 = 36$.

3. Ⓔ When a picture is described, but not shown, draw it. Then, this "hard" question is cake for us. We know to draw a diagram, and we know what to do when we see an isosceles triangle.

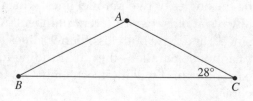

Since $AB = AC$, we know that base angles $\angle B$ and $\angle C$ are equal, and therefore angle B also equals 28. Then since the three angles of the triangle must add up to 180, angle $A = 180 - 28 - 28 = 124$.

4. Ⓗ This question is rated "hard," but easy with our Skills!

Area = 0.5(base)(height)
   72 = 0.5(12)(x)

So $x = 12$ and the triangle is isosceles, which we know means that the base angles are equal. Since it's a right triangle, the base angles must add up to 90, since 180 − the right angle = 90 to be shared by the base angles. Since they are equal, they are 90/2 = 45° each.

Skill 8 (page 27)

FOIL

1. Ⓓ $(2x - 2)(x + 5)$ means multiply $2x$ into the second parenthesis, multiply −2 into the second parenthesis, and then collect like terms:

$2x(x + 5) = 2x^2 + 10x$	multiply $2x$ into the second parenthesis
$-2(x + 5) = -2x - 10$	multiply −2 into the second parenthesis
$2x^2 \underline{+ 10x + -2x} - 10$	collect like terms: $10x + (-2x) = 8x$
$2x^2 + 8x - 10$	

2. Ⓖ $(3m - 2)(5m + 3)$ means multiply $3m$ into the second parenthesis, multiply −2 into the second parenthesis, and then collect like terms:

$3m(5m + 3) = 15m^2 + 9m$	multiply $3m$ into the second parenthesis
$-2(5m + 3) = -10m - 6$	multiply −2 into the second parenthesis
$15m^2 \underline{+ 9m + (-10m)} - 6$	collect like terms: $9m - 10m = -1m$
$15m^2 - m - 6$	

3. Ⓐ Area of a square = side². So $(4x + 3)^2 = (4x + 3)(4x + 3) = 16x^2 + 12x + 12x + 9 = 16x^2 + 24x + 9$.

4. **H** "Product" means multiply. When we multiply the two parentheses, $4x^7$ is the very first term of the solution. "Coefficient" is a fancy term for the number in front of the letter, so for the term $4x^7$, 4 is the coefficient of x^7.

5. **C** To solve this question, you could "Use the Answers" (Skill 1) or you could set up the algebraic expressions, multiply, and solve:

$$(x + 1)(x) = 90$$
$$x^2 + x - 90 = 0$$
$$(x - 9)(x + 10) = 0$$
$$x = 9$$

Now, remember that we solved for the width, and the question asks for the length, which is 1 foot longer than the width. So the answer is C. Careless Error Buster: Remember to ask yourself, "Self, did I finish the question?"

Skill 9 (page 29)

Math Vocab

1. **D** You can do the algebra and solve for x, or you can just "Use the Answers." Most people know that since $(2)(5^2) = 50$, there is at least one answer to this question. But $(2)(-5)^2 = 50$ also. So there are two possible answers to the question, 5 and −5. This is a great reminder to try all the answers and not just pick the first one that works.

2. **J** Lots of students don't know these terms. Since you know, it's easy. Just list the possibilities—7, 9, 11, 13, 15—or "Use the Answers." Either way, as long as you don't get intimidated—as long as you realize that you know the vocab—it's easy! $9 + 11 + 13 = 33$. Make sure to underline the vocab terms and choose the **largest** of the three integers.

3. **C** This question intimidates many people. I know so many students who would immediately say, "I have no chance," and move on. I think that this is something the ACT is testing: can you be intimidated? Shout with me now, "NO WAY!" So once we're done with that, this is easy. Just list a bunch of prime numbers: 2, 3,

5, 7, 9, 11, 13, 17, 19, 23, 29, 31 and look for two in a row (consecutive) that add up to 42. $19 + 23 = 42$. Cake, if you stay relaxed, focused, and confident.

4. **G** Probably best to just find the 2 numbers. If that seems difficult, "Use the Answers" as hints; that is, try −15 and numbers close to −15. The two numbers are indeed −15 and −17. Choice G is a trick that the ACT tried and that we can totally predict. −15 might seem less than −17, since 15 is less than 17, but the bigger the negative number, the less it really is; like the more you owe, the bigger the debt. This sort of trick is discussed more in Skill 46. All other choices are justifiable. This is a great question to use the process of elimination—eliminate choices that you are sure are incorrect (in this case the ones that you are sure are justified) and then choose the best from what's left.

5. **A** Make a list of the integers between 40 and 46: 41, 42, 43, 44, 45. Cross out any consecutive odd prime numbers: 4̶1̶, 42, 4̶3̶, 44, 45. So 42, 44, and 45 are the integers between 40 and 46 that are not consecutive odd prime numbers.

Skill 10 (page 31)

More Math Vocab

1. **C** List the factors of 60, or better yet "Use the Answers." Just divide 60 by the numbers in each answer choice. If the numbers go into 60 evenly (without a remainder), then they are factors. We want the most complete list, the longest list that works.

2. **H** This question tests if you know the terms "odd" and "factor." Once you do, this is a simple "Use the Answers" type of question. Try each answer. Choice A does not work because 2 is not an odd number. Then, to test each of the other choices, simply divide 140 by the number. If it goes in evenly, which means you get an integer (no decimal or fraction), then it is a factor. For example, $140 \div 55 = 2.54$, and

therefore 55 is not a factor of 140. Choice H is correct since $140 \div 35 = 4$.

3. **C** This question simply asks you to factor 50 and count how many of the factors happen to be prime numbers. To factor a number, make a list of pairs, as shown below.

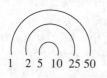

1 2 5 10 25 50

When you reach 5×10, you know that you have them all since nothing between them multiplies to 50. Finding factors systematically like this is far better than randomly jotting down numbers. This is true for any ACT math question. Systematic and organized is better than scattered and random. It avoids careless errors, allows you to look back at your work, and helps you make the leap to the next step needed on a complicated question. So the prime numbers of our list are 2 and 5. Remember that 1 is not a prime number and that 2 is the only even prime number.

4. **K** "Use the Answers" is such a great strategy. Even for this "hard" question, just use the answers. Try each answer choice for a, and see which one could work. Choices F, G, H, and J, when plugged in for a, produce too large a number and would allow a number larger than 36 to be the greatest common factor. For example, when we try the number 9 for a, we get 9^3b and 9^2b^2. The greatest common factor of these numbers will be $81b$, which is greater than 36. Plugging 3 in for a, we get $27b$ and $9b^2$, which could have a greatest common factor of 36.

5. **D** To make this question less theoretical, just try different prime numbers for a, b, and c. We call this "Make It Real;" I'll discuss it more in Skill 18. Let's say $a = 3$, $b = 5$, and $c = 7$. Then $2abc = (2)(3)(5)(7) = 210$. The long way to do this question is to list the factors of 210: 1, 2, 3, 5, 7, 6, 10, 14, 15, 21, 30, 35, 42, 70, 105, 210. The shortcut is to notice that since 2, a, b, and c are prime, the factors will be 1, 2, a, b, c, $2a$, $2b$, $2c$, ab, ac, bc, $2ab$, $2ac$, $2bc$, abc, and $2abc$.

Skill 11 (page 33)

Multiples Vocab

Answer to Brian's Math Magic Trick #1—Multiples: You said carrot. How did I guess your vegetable? Check out my website for answers: www.BrianLeaf.com/carrot

1. **D** This question could be difficult, but "Use the Answers" makes it easy! Just test each answer choice by dividing it by 10, 12, and 15. The lowest choice that is divisible by 10, 12, and 15 is 60. Notice that several other answers work also, but 60 is the lowest. Make sure to try all choices.

2. **G** "Use the Answers."

  F 2 and 62—lcm = 62
  G 9 and 14—lcm = 126
  H 6 and 22—lcm = 66
  J 3 and 42—lcm = 42
  K 5 and 25—lcm = 25

3. **D** "Use the Answers" and it's easy! Just try each answer choice by dividing it by 60, 50, and 80. Choices A, B, and C don't even divide evenly by the three numbers—they are not multiples for all three. Choice D works, all three numbers divide evenly (without a decimal or fraction) into 1200. Choice E works also, but D is the **lowest** answer.

4. **G** "Use the Answers" and it's easy! Just try each answer choice by dividing it by 3, 4, 5, and 6, and then make sure there's an m and a p in there. Choice F doesn't divide evenly by 4. And choices H and K have more m's or p's than we need—we only need one of each. Choice J works, but G is the **lowest** answer. If that seems confusing, you can also "Make It Real" (Skill 18). Just choose easy numbers for m and p, such as 2 and 3, and then test the answers to find the least common multiple. Nice!

5. **B** This "superhard" question is easy for us. The hardest part for most kids is knowing what to do to solve it, i.e. recognizing it as a least common multiple question. But we know that

every ACT has one, and we are watching for it, so we identify it and know what to do. Then just "Use the Answers," although you have to take the two peanuts off before you test each choice to see if it is the least common multiple. Choice B is correct since $362 - 2 = 360$ which is the least common multiple of 8, 9, and 10.

Skill 12 (page 35)

Fancy Graphing Vocab

1. **E** Don't be intimidated by "standard (x, y) coordinate plane," it's just the usual grid we use. Same for "quadrants." I did not include this term in the vocab list because the ACT defines it in the question. The quadrants are just the four areas of the coordinate plane. If x and y have the same signs, then they are either both positive or both negative, and must be in quadrant I, where both are positive, or quadrant III, where both are negative.

2. **J** "Tangent" means that something intersects only one point. So each circle touches an axis at one point. Once you know that, draw a few sketches. Once a sketch shows an answer could work, mark it as a possibility. The two circles could intersect at 0, 1, or 2 points.

3. **A** When you know the vocab, this question is easy points! There are only two things that can make $\dfrac{(x-4)^3}{x^2}$ undefined: a zero on the bottom of the fraction or a negative in a square root. There is no square root in the expression, so that's irrelevant. But zero plugged in for x will give a zero on the bottom of the fraction and is illegal. The math police would arrive within minutes. So the answer is A. As with so many ACT questions, you could also just "Use the Answers" and try the choices for x in your calculator. The ones that don't work, that are "undefined," will give you an error message, or your calculator will implode, but either way you'll know you've got the answer!

Skill 13 (page 37)

Slaloming Slope I

1. **E** Plug the two points into the slope equation.
$$\frac{y_1 - y_2}{x_1 - x_2} = \frac{-4 - 6}{-4 - (-5)} = \frac{-10}{1} = -10$$

2. **F** This question is exactly like question 1. You are given 2 points, and you simply plug them into the slope formula. The only twist is that the question refers to the origin as one of the two points, so you have to know that origin means $(0, 0)$, which is just vocab. Know it (which you do now) and you get it correct!
$$\frac{y_1 - y_2}{x_1 - x_2} = \frac{-6 - 0}{2 - 0} = \frac{-6}{2} = -3$$

3. **D** Plug the two points into the slope equation.
$$\frac{y_1 - y_2}{x_1 - x_2} = \frac{m - (-5)}{2 - (-2)} = \frac{m + 5}{4} = \frac{1}{4}.$$ Once you have
$\dfrac{m + 5}{4} = \dfrac{1}{4}$, you can simply "Use the Answers" and try each answer choice to see which one makes the equation work. You could also cross-multiply and solve for m (see Skill 21).

4. **J** Average rate of score increase is the rate of change, which is just another way that the ACT likes to say "slope." So choose two points such as $(20, 20)$ and $(40, 30)$ and use the slope formula to get $\dfrac{y_1 - y_2}{x_1 - x_2} = \dfrac{20 - 30}{20 - 40} = \dfrac{-10}{-20} = \dfrac{1}{2}.$

Skill 14 (page 39)

Slaloming Slope II

1. **E** Solve the given equation for y.

$3y - 9x = 5$ add $9x$ to both sides
$3y = 9x + 5$ divide both sides by 3
$y = 3x + \dfrac{5}{3}$

Once the equation is solved for y, the slope is the number in front of the x, or 3 in this case.

2. **G** Solve the given equation for y.

$2x + 7y = -14$ subtract $2x$ from both sides
$7y = -2x - 14$ divide both sides by 7
$y = -\dfrac{2}{7}x + (-2)$

Once the equation is solved for y, the slope is the number in front of the x, or $-\frac{2}{7}$ in this case.

3. **C** Slope-intercept form means solve the equation for y.

$6x - 2y - 4 = 0$ add 4 to both sides
$6x - 2y = 4$ subtract $6x$ from both sides
$-2y = -6x + 4$ divide both sides by -2
$y = 3x + -2$

4. **J** Plug the two points into the slope equation.
$\frac{-3 - (-2)}{0 - (-2)} = \frac{-1}{2} = -\frac{1}{2}$. Since the line we want is perpendicular, take the negative reciprocal, $+2$.

5. **D** The origin means $(0, 0)$. First solve for b, plugging the first two points into the slope equation. Then use the value of b in the slope formula for the second pair of points, and solve for a. Because the lines are parallel, the second pair also has a slope equal to 2.

Skill 15 (page 41)

Using Charts and Graphs

1. **C** Stem and leaf plots don't come up that often, but when they do, most kids are like "What the . . . !" That's why I included one here. If you know how to read them, they're easy. They are just a way of listing numbers. Even if you've never seen one before, the "note" tells you how to read it. The "note" tells you that the number "12" would be listed as a "1" in the first column and a "2" in the second column. So this question really is just an easy "mean" question. (Remember from Skill 4 that "mean" is just another word for "average.") So make a list of the 14 numbers, add them up, and divide by 14 to get the average.

2. **H** Many kids get scared off by a question like this because they can't get an exact number from the bar graph. That's okay, no one can. You are supposed to estimate; it's what they want you to do. So estimate the money raised on day 2. It looks a little higher than halfway between $5000 and $10,000, say $8000. Then

this amount divided by the total ($24,500) gives the percent raised on day 2:
$\frac{8000}{24,500} = 0.327 \approx 33\%$. Notice that the answer choices are far enough apart that even an estimate allows us to identify the right answer.

Skill 16 (page 43)

The Only Function Questions on the ACT

1. **D** Simply plug 3 in for x in the equation.

$f(3) = 2(3)^2 + 12 = 18 + 12 = 30$

2. **K** Simply plug $(4, 8)$ in for x and y in the equation.
In other words, plug in 4 for x and 8 for y.

$f(x, y) = 2x - y(4 - x)$
$f(4, 8) = 2(4) - (8)(4 - 4) = 8 - 0 = 8$

3. **B** This is a very typical ACT question, appearing on nearly every ACT. If you got this question wrong and you spend the time to master it, you will gain points! In this question, you are given a value for the number of days, which is d, and you have to determine $F(d)$. Simply plug in the value that you are given (39) for d, and use basic algebra to solve for $F(d)$. If this is confusing, redo this question over and over until you can teach it to a friend. Then do that! This is a great party question. Next time you go to a party, bring this one, people love it. OK, I'm kidding.

$F(d) = 14d + 50$
$F(d) = 14(39) + 50$
$F(d) = 546 + 50$
$F(d) = 596$

Careless Error Buster: Remember to finish the question! Don't choose choice A, 546!

4. **J** This question is just like the "easies," except now we are plugging in variables instead of a number for x. No big deal. Just stay focused and don't get thrown, and it's easy.

$$f(x + k) = 2(x + k)^2 + 3(x + k) - 4$$

plug $(x + k)$ in for each x

$$= 2(x + k)^2 + 3x + 3k - 4$$

multiply $3(x + k)$

That's it. That's all they wanted. The answer is choice J.

5. **E** This question is just like the "easies" except now we are plugging in $h(x)$ instead of a number for x. $h(x) = -x^4$. So $h(h(x))$ means plug $-x^4$, which is a synonym for $h(x)$, in for x:

$$h(h(x)) = -(-x^4)^4 = -x^{16}$$

That's it. That's all they wanted. The answer is choice E.

Skill 17 (page 45)

Midpoint and Distance Formulas

1. **B** A midpoint is halfway between the two points, really just the average. That's what the midpoint formula gives us: $\frac{-14 + 2}{2} = -6$

2. **K** A midpoint is halfway between the two points, really just the average. That's what the midpoint formula gives us: $\left(\frac{-2 + 8}{2}, \frac{4 + 10}{2}\right) = (3, 7)$

3. **E** Use the distance formula:

$$\sqrt{(x - x)^2 + (y - y)^2} =$$
$$\sqrt{(-4 - 1)^2 + [2 - (-10)]^2} = \sqrt{(-5)^2 + (12)^2} =$$
$$\sqrt{25 + 144} = \sqrt{169} = 13.$$

4. **F** The center of the circle is in the middle of the diameter—the midpoint of the diameter. Midpoint $= \left(\frac{-2 + 6}{2}, \frac{10 - 4}{2}\right) = (2, 3)$

5. **C** This word problem is just asking for the distance between the two restaurants. Use the distance formula: $\sqrt{(x - x)^2 + (y - y)^2} =$

$$\sqrt{(17 - 12)^2 + (15 - 5)^2} = \sqrt{(5)^2 + (10)^2} =$$
$$\sqrt{25 + 100} = \sqrt{125}$$

Skill 18 (page 47)

Make It Real

1. **B** This question intimidates some students; it seems too theoretical. But "Make It Real" makes it so easy. Just choose a real number for p (remember that the question says that the number that you choose has to be odd). Let's say $p = 5$. Then try $p = 5$ in all the answer choices to see which one does what the question asks, which one gives an odd number. The answer is B since $2(5) - 1 = 9$.

2. **J** First, "product" means "multiply." We will review this more in Skill 27. "Make It Real," let's say, $m = -3$ and $n = 3$. Now it's easy, the product (multiplication) of -3 and 3 is -9. Choice J looks good, but to be safe, try a few more possibilities. After you've done three, you can be sure that your answer is correct. And by then you might realize that "Of course, a positive times a negative is always a negative."

3. **D** Choose a number for b, let's say $b = 5$ (notice the question states that b can be any real number besides 3 and 7). Now it's easy, solve:
$$\frac{(3 - 5)(5 - 7)}{(5 - 3)(5 - 7)} = \frac{(-2)(-2)}{(2)(-2)} = \frac{4}{-4} = -1$$
Skill 12 review: the value of b cannot equal 3 or 7 because either number would give us a 0 on the bottom of the fraction, which is illegal in math and would cause the equation to be "undefined" and would cause lots of important folks to get very upset.

4. **H** We are into the "hards" now. But "Make It Real" will knock these down a notch to mediums or even easies. Let's say $k = 100$ and $q = 20$. So, of the 100 kids in a class, 20 have NOT seen *Monty Python and the Holy Grail*. Once we've "Made It Real," it becomes obvious that if 20 have NOT seen it, then 80 have seen it. (When you are using "Make It Real" with percents, use 100 when you can, it makes questions really easy.) Now, based on $k = 100$ and $q = 20$, we want an answer of 80. So try each answer choice, using $k = 100$ and

$q = 20$ and find the one answer choice that equals 80.

Choice H is correct because $\frac{(100 - q)k}{100} = \frac{(100 - 20)100}{100} = 80$.

If two choices work, just choose new "Make It Real" numbers and have a tiebreaker.

People think that this might take forever to do, but try it; it takes like 30 seconds.

5. **E** Let's say $m = 3$ and $n = 5$. Then $-3m - 2n - 4 = -3(3) - 2(5) - 4 = -23$. The question asks what happens when m increases by 2 and n decreases by 1, so let's solve for $m = 5$ and $n = 4$. Now, $-3m - 2n - 4 = -3(5) - 2(4) - 4 = -27$. So the result changed from -23 to -27; it decreased by 4.

Skill 19 (page 49)

Perimeter, Area, Volume

1. **E** To find perimeter, just add up the sides. To avoid a careless error and to make it clear what to do next, sketch a diagram. Since all sides of a square are equal, perimeter = 4(side) = 4(75) = 300.

2. **J** The volume of a rectangular solid, i.e., a box, equals (length)(width)(height). Just fill the info that we know into the formula, and solve for the variable that we are missing. This is a great word problem strategy. It solves 90% of all formula-related word problems. Just "fill in what we are given and solve for whatever is still a variable." So

   Volume = (length)(width)(height)
   $10,000 = (75)(110)(h)$
   $10,000 = 8250h$
   $1.21 = h$

3. **D** The area of a parallelogram = (base)(height). So

   Area = (base)(height)
   Area = (40)(15) = 600

   The key here is to memorize the parallelogram formula and to remember that 15 is the height,

not 17. Height is always at a right angle, like when the nurse at school is measuring your height and says, "Stand up straight, no slouching." Also, remember to add the two parts of the base, $8 + 32 = 40$.

4. **K** The word "circumscribed" throws some kids. Lot's of kids say, "I don't know that word, I'll just guess and move on." But in this book you'll learn all that you need for the ACT. When they throw a tough word at you that we have not discussed in this book, they'll define it in the question or you won't even need it. Here we don't need the word; you can cross it out. It's totally implicit in the diagram. "Circumscribed" just means "drawn around." So the square is drawn around the circle, which is obvious from the diagram anyway! So, don't get intimidated. Stay confident and focused. So the area of the square = (length)(width). And for a square, length and width are equal. You can see in the figure that the circle has a radius of 9 and a diameter of 18, and you can see that the diameter is the same as the length of a side of the square, so its area = (18)(18) = 324.

5. **C** The ACT loves these. The key here is to realize that all the vertical lengths of the steps add up to 7, and all the horizontal lengths of the steps add up to 10. Now that you know this trick, it's easy! So perimeter = 34.

Skill 20 (page 51)

Donuts

1. **D** When a picture is described and not shown, draw it. That helps you see what to do next, and it helps avoid careless error. The area of the remaining portion equals the area of the wall minus the area of the windows:

   Area wall − 3(area window) = area remaining portion
   $(7)(10) - 3(2)(2) = $ area remaining portion
   $70 - 12 = $ area remaining portion
   $58 = $ area remaining portion

2. **K** The word "circumscribed" throws some kids. But, in this book you'll learn all that you need for the ACT. When they throw a tough word at you that we have not discussed in this book, they'll define it in the question or you won't even need it. Here we don't need the word; you can cross it out. It's totally implicit in the diagram. "Circumscribed" just means "drawn around." So the square is drawn around the circle, which is obvious from the diagram anyway! So don't get intimidated. Stay confident and focused. So, the area around the circle equals the area of the square minus the area of the circle. "Mmmm . . . donuts!"

Area square − area circle = area around circle
$(18)(18) − \pi(9)^2 = 324 − 81(\pi)$

3. **C** Some kids see this and say, "I don't know the formula for the area of a trapezoid, so I can't do it." It's true that you could solve this question with the trapezoid formula, but you can also use donuts. They give you a back door on purpose. Don't get intimidated. Stay with it and look for another way! The area of the shaded trapezoid equals the area of the rectangle minus the areas of the two triangles:

Area trapezoid = area rectangle − area of triangles
Area trapezoid = $48 − 0.5(6)(4) − 0.5(3)(4)$
Area trapezoid = $48 − 12 − 6 = 30$

4. **H** First, always label all the info from the question into the diagram. That helps you see what to do next and to avoid careless errors. The area of the shaded region equals the area of the big square minus the area of the small square. We can do that with some special right triangles or Pythagorean theorems (Skills 25 and 24), but we can also just calculate the area of a shaded triangle and multiply by 4.

Area shaded = 4(area triangle)
Area shaded = $4(0.5(4)(4))$
Area shaded = 32

Skill 21 (page 53)

Baking Granola Bars . . . Ratios, Proportions, and Cross-Multiplying

1. **B** Label info from the question into the diagram. That helps you see what to do next and avoid careless errors. Once you label the diagram, you see that when $MN : MO$ is 2 : 3, then $MN : NO$ is 2 : 1. This is a classic ratio question where you must switch the ratio from part : whole to part : part.

2. **F** Set up the proportion. Cross-multiply, and it's easy. Pretty much anytime you see a proportion, cross-multiply. $\frac{60}{15} = \frac{x}{3}$. So $(60)(3) = 15x \Rightarrow 180 = 15x \Rightarrow 12 = x$.

3. **E** Set up a proportion for blueprint length and actual length: $\frac{\frac{1}{8}}{1} = \frac{2\frac{1}{8}}{x}$. Cross-multiply to get $\frac{1}{8}x = 2\frac{1}{8}$. Divide both sides by $\frac{1}{8}$ to get $x = 17$.

If you don't like fractions, no problem, just use your calculator to convert them to decimals. Lots of kids skip this one to avoid fractions. But you can just convert them right at the start and never have to deal with fractions at all! I love that strategy! So $x = 17$ is one length of the room, and the other is

$\frac{\frac{1}{8}}{1} = \frac{2}{x}$. Cross-multiply to get $\frac{1}{8}x = 2$. Divide both sides by $\frac{1}{8}$ to get $x = 16$.

Thus, the room is 17 by 16, and its area = $(17)(16) = 272$.

4. **J** The ratio of x to y is $\frac{2}{5}$, and the ratio of z to y is $\frac{3}{4}$. We want a ratio between x and z, so we need to get the y's (the bottoms of the two fractions) to match up, because then we could drop the y's and just compare x to z. The two ratios $\frac{2}{5}$ and $\frac{3}{4}$ could match if their bottoms

were 20 (the lowest common denominator of 5 and 4): $\frac{8}{20}$ and $\frac{15}{20}$. Then we can compare $x:z$ as 8:15.

Skill 22 (page 56)

Intimidation and Easy Matrices

1. **A** Notice that this question asks for the determinant, which I didn't teach you. That's because the ACT defines it in the question, and you just have to follow the directions that are given. You could get this one even if you never heard of a matrix, as long as you didn't make the connection and freak out. Sometimes ignorance is bliss. On the ACT, always follow the directions that are given in a question; they give you lots of good info. So plug $m = -2$ and $n = 3$ into the matrix and follow the directions for the determinant to get $(3(-2))(3(3)) - (2(-2))(4(3)) = -54 - (-48) = -6$

2. **K** These matrices are just charts showing data. Take a moment to understand the charts, and then the question is easy. In fact, if we dropped the word "matrix" from this question, fewer people would be intimidated, and it'd be ranked easier. So don't ever get intimidated by the word "matrix". It's just a chart of data. Angino won the election with 50 votes. The part of his votes that came from juniors was 0.2, so multiply $50 \times 0.2 = 10$ votes from juniors.

3. **C** Again, just a chart of data, like Skill 15. Take a moment to understand the chart and then it's easy. A 1 tells us that the lights in that section are on, and a 0 tells us that they are off. So choice C is correct, the lights in sections A, C, and H are on.

4. **H** To add matrices, just add the terms that are in the same spot in each matrix. So
$$\begin{bmatrix} p & q \\ r & s \end{bmatrix} + \begin{bmatrix} w & x \\ y & z \end{bmatrix} = \begin{bmatrix} p+w & q+x \\ r+y & s+z \end{bmatrix}$$

Skill 23 (page 59)

Art Class

1. **D** When a picture is described, draw it. Nobody can just visualize this and get an answer. Don't be a hero here; draw a diagram.

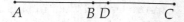

Since $AB = 8$ and $AC = 8 + 8 = 16$, AD must be between 8 and 16, so only choice D works.

2. **H** When a picture is described, draw it. Not even your math teacher, the one who walks around running his finger along the walls muttering math calculations to himself, could solve this one in his head. Draw a somewhat neat diagram, and it's easy! Plus, using a diagram avoids careless errors. Notice that the answer choices are far enough apart that a fairly neat diagram will show you which answer choice is correct!

3. **C** First, label the info from the question into the diagram. Then ask yourself if the diagram seems accurate. Yes, it does, because DB does look like half of EB, and DC looks like a little less than DB, so the numbers work. Then if the geometry jumps out at you, by all means do it, and after our geometry Skills coming up, they probably will, but you can also "Use the Diagram!" Since EA looks longer than CD but shorter than EB, it must be between 6 and 14. That only leaves choices C and D. Pretty cool, huh? Now, is it C or D? Well, EA is only slightly longer than DC, not almost double, so I'd go with choice C, which is correct! Sometimes "Use the Diagram" is enough to get the right answer; sometimes it's just a plan B to narrow it down, when you can't decide how to get it with geometry Skills.

4. **G** This is a great "Use the Diagram" question! First, label the info from the question into the diagram. To do this question the "math class way" would require using both the midpoint formula and the distance formula, which we

certainly know how to do from Skill 17. But we can save a ton of time by just looking at the diagram and the answer choices. Connect point P to Q. Then sketch a point halfway between them, the midpoint of the segment. Then connect point M to that midpoint. Voilà! Now, how long does that segment look, based on the fact that MP is 4? It looks about the same, actually. Then look at the choices. Wait, none of them say 4. Do we panic? No, convert them into decimals. Aha! Choice G is about 4.

Skill 24 (page 61)

The Mighty Pythagoras

1. **C** Great review of Skill 23; when a picture is described, draw it. So first draw a diagram. Then since you see a right angle, you'll probably need our good pal Pythagoras and his $a^2 + b^2 = c^2$. In this case, 6 and 8 are the a and b, and c is what we are looking for. How do I know? The diagram shows me; c is always the side opposite the right angle, it's as simple as that! So

    $6^2 + 8^2 = c^2$.
    $36 + 64 = c^2$.
    $100 = c^2$
    $10 = c$

2. **J** Another great review of Skill 23; when a picture is described, draw it. Man, that one strategy helps sooo much. Once you have a diagram, you can see that we have a right triangle situation here. Pythagoras to the rescue:

    $27^2 + 36^2 = c^2$
    $729 + 1296 = c^2$
    $2025 = c^2$
    $45 = c$

3. **A** Lots of kids get tripped up by "coordinate units." Remember, if something throws you, don't panic and don't just guess and move on. Basically, just ignore the thing that throws you, you probably don't need it—you can work without it. The questions are designed for that.

Just cross off "coordinate units" if it bothers you! (It just means units on the coordinate plane.) Then label info from the question into the diagram—another huge strategy. Drawing the segments shows you that we have a right triangle. And remember that hypotenuse means the longest segment of the triangle, the one opposite the right angle. Using the graph and the coordinates, determine the lengths of sides and plug them into $a^2 + b^2 = c^2$. According to the diagram, the lengths of the short sides of the triangle are 2 and 6. So

$2^2 + 6^2 = c^2$
$4 + 36 = c^2$
$40 = c^2$
$\sqrt{40} = c$

Hey, that's not an answer! Relax, and please do not hurt me. Just translate $\sqrt{40}$ into a decimal, and then translate the answers and choose the closest one. $\sqrt{40} \approx 6.32$, and $2\sqrt{10} \approx 6.32$. Bingo, choice A works perfectly. By the way, as a third-degree master of the distance formula, you could, of course, have used it here as another way to find the answer.

4. **F** Fun question. Fun, because it's easy with our strategies! Anytime you see a right triangle on the ACT, try $a^2 + b^2 = c^2$. Don't even think; as soon as you see a right triangle, try $a^2 + b^2 = c^2$.

    $m^2 + (3m)^2 = c^2$
    $m^2 + 9m^2 = c^2$
    $10m^2 = c^2$
    $\sqrt{10}\,m = c$

This is a great preview of exponents (Skill 30) and avoiding careless errors (Skill 44).

Skill 25 (page 63)

Special Right Triangles

1. **E** This question is rated "medium" only because most kids do not know special right triangles. If you take five minutes right now and memorize them, just the info on the page for Skill 25, then this "medium" becomes easy

and you gain points! So once you know about special right triangles, E is obviously the answer. It follows the pattern: x, $x\sqrt{3}$, $2x$. If, for some crazy reason, you have not memorized the pattern, first of all shame on you, you silly slacker, but second of all, you could draw a diagram for each choice and see which one looks 30-60-90ish.

2. **J** Great review. You can certainly use special right triangles here, but you could also just "Use the Diagram!" Remember "Use the Diagram"? First you look at it and ask if it seems drawn to scale. Yes, they look like squares, and M, N, O, and P look like midpoints—they are in the middle of each segment. So we can use the picture. If WX is 8, clearly NO is less than 8, perhaps 5 or 6. So the perimeter of $MNOP$ is about 5.5(4) = 22. Convert the answer choices to decimals and choose the best answer. Awesome strategy! The correct answer is J; it's the only one that's even close. Another "medium" made easy! It's so funny that with a $\sqrt{2}$ in it, an answer choice seems like a foreign language, but translated to decimals, it's so obvious! We can also solve this with special right triangles. Since $WXYZ$ and $MNOP$ are squares, triangle NXO is a 45-45-90 triangle with sides 4, 4, $4\sqrt{2}$, and the perimeter of $MNOP$ is $4\sqrt{2} + 4\sqrt{2} + 4\sqrt{2} + 4\sqrt{2} = 16\sqrt{2}$.

3. **D** As soon as you see a right triangle with a 60° angle, use the special right triangle 30, 60, 90. The hypotenuse is 12, so the smaller leg must be 6, and the other leg is $6\sqrt{3}$. A triangle congruent to the one shown would also have a longer leg of $6\sqrt{3}$, since they are congruent.

4. **G** An isosceles right triangle is a right triangle where two sides are the same length. When two sides are equal (great review of Skill 7), their two opposite angles are also equal. Therefore, we have a 90-45-45 triangle. Since the two equal sides add up to 9, each is 4.5, and the longest side is $4.5\sqrt{2}$.

5. **D** The sides of a 30-60-90 triangle follow the pattern x, $x\sqrt{3}$, $2x$. So if the shortest and longest sides add up to 12, then $x + 2x = 12$. So $3x = 12$ and $x = 4$. Then the other side measures $4\sqrt{3}$. Nice!

Skill 26 (page 65)

Dr. Evil and Mini-Me

1. **E** To get the perimeter of $\triangle XYZ$, we just add up the lengths of the sides. We are missing one side, which we will get by using similar triangles. How did I know? Because the question told me. If it mentions similar triangles, you will use them, guaranteed. That's how the ACT works. So when it says similar triangles, set up a proportion and solve for a side. The letters of the triangles will be written in corresponding order, so XY corresponds to PQ, since they are the first two letters in the order that each triangle is written, $\triangle XYZ$ and $\triangle PQR$. Since similar triangles have proportional sides, to solve for XY, set up the proportion:

$$\frac{\text{Middle}}{\text{Middle}} = \frac{\text{shortest}}{\text{shortest}}$$
$$\frac{XY}{8} = \frac{18}{6}$$

Cross-multiply to get $144 = 6XY$. Divide both sides by 6 to get $24 = XY$. Thus, the perimeter of $\triangle XYZ = 24 + 18 + 27 = 69$.

2. **H** When a picture is described, draw it. Triangle ABC is isosceles, with $AC = BC$. Similar triangles have proportional sides, and their letters are written in corresponding order. So since MN corresponds to AB, and MO corresponds to AC, we can set up the proportion: $\frac{15}{5} = \frac{MO}{4}$

Cross-multiply to get $60 = 5(MO)$. Divide by 5 to get $MO = 12$.

3. **B** When a picture is described, draw it. This shows you that we have two triangles, which are similar. Set up a proportion:

$\frac{\text{person}}{\text{shadow}} = \frac{\text{lamppost}}{\text{shadow}} \Rightarrow \frac{2}{4} = \frac{x}{15}$. When you see a

proportion, cross-multiply: $30 = 4x$. And divide by 4 to get 7.5.

4. **G** When a picture is described, draw it. We want the perimeter of ΔXYZ, and we only have the longest side, which must be XZ since it corresponds to the longest side of ΔPQR (we can tell by the order of the letters in ΔXYZ and ΔPQR). So we set up proportions to solve for each side of ΔXYZ:

$\frac{\text{Shortest}}{\text{Shortest}} = \frac{\text{longest}}{\text{longest}} \Rightarrow \frac{XY}{6} = \frac{35}{14} \Rightarrow 14XY = 210 \Rightarrow$

$XY = 15$

$\frac{\text{Middle}}{\text{Middle}} - \frac{\text{longest}}{\text{longest}} \Rightarrow \frac{YZ}{8} = \frac{35}{14} \Rightarrow 14YZ = 280 \Rightarrow$

$YZ = 20$

So the perimeter of $\Delta XYZ = 20 + 15 + 35 = 70$.

5. **C** If two triangles have two equal angles, then they are similar. Basically, on the ACT if two triangles seem similar, they probably are. So since they are similar, their sides are proportional, smallest to smallest and largest to largest. Let's say $BY = x$, then $ZY = 450 + x$. Set up the proportions and cross-multiply:

$\frac{600}{440} = \frac{450 + x}{x}$

$600x = (440)(450) + 440x$
$160x = 198,000$
$x = 1237.5$

Skill 27 (page 69)

Is Means Equals—Translation

1. **B** Simply translate, word for word, and convert fractions to decimals.

"what" means x.
"is" means =.
"of" means multiply.
So $x = (0.125)(0.30)(6000) = 225$.

2. **G** Translate. The slower plane is represented by s. The faster plane is 20 more than $3s$, which is $3s + 20$.

3. **C** Simply translate, word for word:

"is" means =.
"what" means x.
"of" means multiply.

So $31 = (x)(155)$. Divide both sides by 155 to get $x = 0.20$. Then since the question asks for "what percent," translate 0.20 to 20%.

4. **J** Now 60% of 1000 means (0.60)(1000) which equals 600. And 10% of this 600 means (0.10)(600), which equals 60. So the 10% who will buy the movie equals 60 kids.

5. **C** Simply translate, word for word: $(1.25)(x) = 750$. So $x = 600$, and $(0.65)(600) = 390$.

Skill 28 (page 71)

Arithmetic Word No-Problems

1. **C** Simply translate, word for word. "3.5 pounds of corn at \$0.35 per pound and 1.5 pounds of tomatoes at \$0.74 per pound" means $(3.5)(0.35) + (1.5)(0.74) = 2.335 \approx 2.34$

2. **H** Simply translate. Evan pays a \$65 flat fee plus \$5 for every 3 movies. So his total cost is $65 + 5(492 \div 151\ 3) = \$885$.

3. **D** Translate fractions to decimals (by dividing the top number of the fraction by the bottom number), and then it's easy! Laurie ran 3.8 miles on Tuesday and 4.66 miles on Wednesday, so she ran a total of $3.8 + 4.66 = 8.46$ miles. Choice D is correct, 8.46 is between 8.4 and 8.66.

4. **H** Translate: 3 cases with 12 two-pint containers means $(3)(12)(2) = 72$ pints. The same amount of berries would be in $72 \div 3 = 24$ three-pint containers.

Skill 29 (page 73)

Just Do It!—Springboard

1. **C** Great springboard question. When you see a question like this, ask yourself, "Self, how do these equations relate?" There must be a quick easy way to do this, that's how they design the ACT. That's where springboard comes in. It tells you to factor whenever you can—if you can factor, that's what they want you to do, it's how to get the question correct. When you factor the second equation, you get $z = 6(x - y) - 5$. Since we know that $x - y = 4$, we plug in 4 for the $x - y$ to get $z = 6(4) - 5 = 19$. Nice!

   Moral of the question: Always ask, "How do the two equations relate?"

2. **F** The ACT loves $x^2 - 49$. When you see $x^2 - 49$, factor it to $(x - 7)(x + 7)$. Then the next step becomes clear—springboard helps you see what to do next. You don't even need to see it all in advance, just take it one springboard step at a time. So once you write $(x - 7)(x + 7)$, you see that there is an $(x - 7)$ on top too. Rewrite $(x - 7)^2$ as $(x - 7)(x - 7)$, and reduce/eliminate one of the $(x - 7)$'s:

$$\frac{(x - 7)^2}{x^2 - 49} \Rightarrow \frac{(x - 7)(x - 7)}{(x - 7)(x + 7)} \Rightarrow \frac{x - 7}{x + 7}$$

3. **C** Nine out of ten kids freak out, guess, and move on. But if you translate fractions to decimals, it's so easy! I love this strategy! So

$$\frac{8 + \frac{3}{8}}{1 + \frac{3}{16}} = \frac{8 + 0.375}{1 + 0.1875} = \frac{8.375}{1.1875} = 7.05$$

Then, just convert the answer choices and choose the best fit. Choice C is correct because $\frac{134}{19} = 7.05$.

4. **F** Springboard, baby! Springboard helps you see what to do next. Follow all our springboard rules: When you have two equations like this, ask, "How are they connected; what's the back door?" Just asking the question usually helps you find the connection. Then you notice that

$\frac{a}{b} = 8$ means that $a = 8b$. (When you see a proportion, cross-multiply. It's how the ACT works! It's always the way to get the answer.) Next springboard rule: "When you see $x^2 - y^2$, **factor** it to $(x - y)(x + y)$." Guaranteed, if you see $a^2 - 64b^2$, that's what you need to do. So $a^2 - 64b^2 = (a - 8b)(a + 8b)$. Lastly, plug $8b$ in for a to get $(8b - 8b)(8b + 8b) = 0$, since $8b - 8b = 0$. Nice. Now that was fun.

Skill 30 (page 75)

Beyond Your Dear Aunt Sally:
The Laws of Exponents I

1. **E** Multiply coefficients (numbers in front of the letters) and add exponents. So $2x^2 \cdot 3x^3y \cdot 3x^3y = (2 \cdot 3 \cdot 3)(x^{2+3+3})(y^{1+1}) = 18x^8y^2$. Remember that y really means y^1.

2. **G** When you multiply variables, like m, you add the exponents; and when you divide m's, you subtract the exponents. So, $m^3 \times 2m^5 \times \frac{1}{m^2} = 2m^{3+5-2} = 2m^6$.

3. **D** Springboard review: when you see something that can simplified, do it. $x^3 = 27$ simplifies to $x = 3$, since $3^3 = 27$. Then just plug 3 in for each x: $3^3 + 3^2 = 27 + 9 = 36$.

4. **K** $(-3x^3y^4)^4 = (-3)^4(x^3)^4(y^4)^4 = 81x^{12}y^{16}$. Careless Error Buster: Remember that the (-3) also gets the exponent.

5. **B** 186,000 is in miles per second. To get miles per minute, multiply by 60. And then to get miles per hour, multiply by 60 again. Finally, to get 3 hours, multiply by 3: $186{,}000 \times 60 \times 60 \times 3 = 2{,}008{,}800{,}000$. Then just type each answer choice into your calculator to see which one matches best. Choice B, 2,000,000,000, is closest. A number written in the form 2×10^9 is called scientific notation, and just means 2 with 9 zeros, or 2,000,000,000. So even if a number in scientific notation was too big for your calculator, just remember that it means the first number with as

many zeros as the number in the exponent, like $5 \times 10^4 = 50{,}000$.

6. **H** Since $m^p = x$ and $n^p = y$, then $xy = (m^p)(n^p) = (mn)^p$. Notice that this time I did not add the exponents because m and n are different letters, and we only add exponents when the base is the same.

Skill 31 (page 77)

Far Beyond Your Dear Aunt Sally:
The Laws of Exponents II

1. **C** $2x^2$ and $2x^3$ are not like terms and do not combine. To combine, the variable (x) **and** the exponents would have to match.

2. **F** To multiply $3m^{-2}$ by $2m^{-5}$, we add the exponents, so $(3m^{-2})(2m^{-5}) = 6m^{-7} = \dfrac{6}{m^7}$.

3. **E** Get the common denominator for the fraction exponents, which is 6, and then the top of each converted fraction is the exponent in the answer. If that gave you a headache, just "Make It Real." Convert fractions to decimals ($a^{2/3}b^{1/6}c^{1/2} = a^{0.666}b^{0.1666}c^{0.5}$), and then choose numbers for a, b, and c, such as $a = 3$, $b = 4$, and $c = 5$. Get an answer, and see which answer choice gives the same answer. $(3)^{0.666}(4)^{0.1666}(5)^{0.5} = 5.855$. Choice E is correct, since $\sqrt[6]{(3)^4(4)(5)^3} = 5.86$. "Make It Real" makes "hard" questions easy!

4. **K** There is the long "math class" way to do this "hard" question, and there is the flawless crazy-easy "Use the Answers" way! First, translate fractions to decimals. Then just try each choice in the question and see which one makes the equation work. Choice K is correct, since $\left(\dfrac{3}{4}\right)^{-2} = \sqrt{\left(\dfrac{4}{3}\right)^4} = 1.77$. Nice!

Skill 32 (page 79)

$y = mx + b$

Answer to Brian's Math Magic Trick #2: come on, you should know that there are no elephants in Denmark! To find out how this stunning act of magic works, go to my website: www.BrianLeaf.com/Elephants

1. **B** 25 people are disturbed by each scene, so that number goes next to the x. 45 people were offended right off the bat by the trailer, and this number does not change with the number of disturbing scenes in the movie, so that number goes alone. $y = 25x + 45$.

2. **J** "Use the Answers"! Just choose an (x, y) pair from the table, such as $(8, 5)$, and test each answer. If two choices work, choose another (x, y) pair and do a tiebreaker. Choice J works since $5 = 0.25(8) + 3$.

3. **B** The graph shown crosses the y axis at 2, so it has a "y intercept" of 2, meaning $b = 2$. The slope is positive because the line rises from left to right, so $m > 0$. (Review slope in Skill 13 if needed. If you slacked during the slope chapter and did not do the drills, e-mail me and I'll send a motivational speech your way.) Choice B is the only choice that satisfies both requirements: y intercept of 2 and positive slope.

4. **K** The ACT loves $y = mx + b$, and they are obsessed with fixed costs, which are always the b. Here, we need to add more rows to the table until we get 0 inserts. At zero inserts, the only cost will be the fixed charge, since we won't be paying for any inserts. If we add a row above the current data, for a cost of 33, we have 0 inserts, so the fixed cost is 33.

5. **C** Plug 20 in for h, which represents height. So $20 = 3.4s$, and dividing by 3.4, we get $5.88 = s$, which is between 5 and 6 seconds.

Skill 33 (page 82)

Arrangements

1. **E** Draw a blank for each option. Then write in the number of possibilities that can fill each option. There are 2 slogan options and 3 size options, so $2 \times 3 = 6$. Notice that this question is simple enough to do without the strategy;

you could just picture shirts in piles, like at Old *Navy*, with two different slogans available in S, M, and L, making 6 piles.

2. **J** Draw a blank for each role that needs to be filled by an actor. Then write in the number of actors who can fill the role. Remember that each actor can play only one role, so once someone is assigned, that person can't be used again. Then multiply.

    $5 \times 4 \times 3 \times 2 \times 1 = 120$.

3. **D** Draw a blank for each spot at the burrito bar. Then write in the number of possibilities that can fill each slot. There are 3 options for meat, 3 options for vegetables, and 4 options for salsas, so $3 \times 3 \times 4 = 36$.

4. **K** Draw a blank for each spot on the banner. Fill in the number of possible symbols that can go in each spot. Notice that the three symbols must be different, since the question states that Kyle will use **three** of the symbols. Therefore, they cannot be reused, and we have $5 \times 4 \times 3 = 60$.

5. **C** Draw a blank for each member of the doubles team. Write in the number of possible girls who can fill each slot; remember that once someone is assigned a position, she cannot also play another position. Then multiply. This question has one extra step, since this is a team of two, and it does not matter if it's Jenny and Jill or Jill and Jenny. So we divide our answer by 2 since we will have double-counted each duo. $6 \times 5 = 30 \div 2 = 15$.

Skill 34 (page 85)

SohCahToa!

1. **B** We always want to mark all the info from the question into the diagram. And sometimes the ACT does that for us. Here all info is marked. Thanks, ACT. So don't get caught up in the complex wording of the question. Look at the diagram. It's a straightforward

SohCahToa question. The key to SohCahToa questions is simply to determine which ratio we need for the question. All we have to do is look at the info given. If we are given opposite and hypotenuse, we use sin. If we are given adjacent and hypotenuse, we use cos. And if we are given opposite and adjacent, we use tan. That's it. So based on the 40° angle, we have the adjacent leg and want the opposite leg. So, we use tan:

$$\tan = \frac{\text{opposite}}{\text{adjacent}}$$

$$\tan 40 = \frac{x}{4}$$

$0.839 = \frac{x}{4}$ use your calculator to get the value of tan 40, which is 0.839

$3.36 = x$ multiply both sides by 4

2. **G** Remember that sin, cos, and tan are talking about right triangles, so draw a diagram. Sin represents $\frac{\text{opposite}}{\text{hypotenuse}}$, so since $\sin A = \frac{3}{5}$, we know that the opposite leg of vertex A equals 3 and the hypotenuse equals 5. Whenever you have two sides of a right triangle, use $a^2 + b^2 = c^2$ to get the third side. Great mighty Pythagoras review (Skill 24). So $3^2 + b^2 = 5^2$ and $b = 4$. Therefore, $\tan A = \frac{\text{opposite}}{\text{adjacent}} = \frac{3}{4}$, since the opposite leg equals 3, and the adjacent leg equals 4.

3. **A** $\sin B = \frac{\text{opposite}}{\text{hypotenuse}}$. In the diagram, the leg opposite vertex B equals 5, and the hypotenuse equals 7. So $\sin B = \frac{\text{opposite}}{\text{hypotenuse}} = \frac{5}{7}$.

4. **J** $\cos Z = \frac{\text{adjacent}}{\text{hypotenuse}}$. In the diagram, the leg adjacent to vertex Z equals 4, and the hypotenuse equals 5. So $\cos Z = \frac{\text{adjacent}}{\text{hypotenuse}} = \frac{4}{5} = 0.8$.

Skill 35 (page 87)

Beyond SohCahToa

1. **C** Awesome "Use the Answers" question! Try each answer choice, and use the process of elimination until you find the 1 choice that has values that all work. Choice C is correct since both 45 and 225 yield 1 when plugged in for θ in the expression tan θ. If you've studied trig, you could also do this question the "math class way." Since tan means "opposite over adjacent," tan θ = 1 when opposite = adjacent, and therefore sin = cos. So what are the values for θ where sin = cos? Sin could equal cos when θ is between 0 and 90 or between 180 and 270, because in these regions sin and cos are either both positive or both negative. Use the process of elimination, and only choice C has answers in these two regions.

2. **F** Follow the directions given for the law of sines. The ratio of a side and the sin of its opposite angle is equal for all sides. So $\frac{32}{\sin 74} = \frac{27}{\sin 54} = \frac{x}{\sin 52}$. Perimeter equals the sum of the sides of the triangle, and we already know two of the sides. We can use that ratio to solve for x, the third side. So since $\frac{32}{\sin 74} = \frac{x}{\sin 52}$, we can cross-multiply to get $(x)(\sin 74) = (32)(\sin 52)$; and divide both sides by sin 74 to get $\frac{32 \sin 52}{\sin 74} = x$. Therefore, the perimeter equals $59 + \frac{32 \sin 52}{\sin 74}$.

3. **D** Notice that all of these are "hards." You would only see this Beyond SohCahToa stuff as hards. You can "Use the Answers" here. Graph each answer choice on your calculator, and find the one that matches the graph shown in the question. You'll notice, when you graph these, that they don't make the curved line like in the picture; that's your tipoff that you need to change your calculator to radians mode. That's the only curveball for this particular type of question. You need to change the mode on your

calculator to radians, and you have to remember to switch it back when you're done. Just ask your math teacher to show you this if you've never done it before. Radians are just another way (besides degrees) to measure an angle. Once you know how to do this on your calculator, choice D matches perfectly.

If you've studied trig, you could also do this question the "math class way." The graph shown has been shifted from its usual position. When a graph has been shifted, we can use the equation $y = a \cos b(x - c) - d$, where a tells the altitude of the curve (how high up and down it reaches), $2\pi/b$ tells the period (the length of one repeat), c tells how far left or right the graph was moved from the origin, and d tells how far up or down the graph was moved from the origin. The graph in the question has been shifted down and to the right from the ordinary cos graph. So we want a cos graph with a right and down shift, represented by the c and d. So answer choice D is correct. It is the only one with a number for c and d in the equation $y = a \cos b(x - c) - d$.

4. **K** This question is crazy theoretical, so you can "Make It Real." Just choose positive numbers for b, c, and d and graph the equation on your calculator. Then find the minimum value (the lowest point) of the graph. You'll notice when you graph these that they don't make the usual trig curved graph like the diagram in question 3; that's your tipoff that you need to change your calculator to radians mode. That's the only curveball for this particular type of question. You need to change the mode on your calculator to radians, and you have to remember to switch it back when you're done. Just ask your math teacher to show you this if you've never done it before. Radians are just another way (besides degrees) to measure an angle, and once you know how to do this on your calculator, it's easy.

To "Make It Real," let's say $b = 2$, $c = 3$, and $d = 4$; then the graph shows a minimum value of −5. Using the numbers we choose for b, c, and d in the answer choices, we need the choice that also yields −5. Choice K works, since −1 − 4 = −5.

If you've studied trig, you could also do this question the "math class way." In the equation $y = a \cos b(x - c) - d$, the c tells how far left or right the graph was moved from the origin, and d tells how far up or down the graph was moved from the origin. So this graph will always reach a minimum of −1 − d, since it's normal minimum is −1 and it has been shifted down d units, giving a new minimum of −1 − d.

Skill 36 (page 89)

What You Really Want . . . Probability

1. **B** Probability = $\frac{want}{total}$, so the probability of NOT selecting a green marble is $\frac{4}{12}$. This question requires that we reduce the $\frac{4}{12}$ to $\frac{1}{3}$. No problem. Stop looking at me like that, reducing is easy, just divide the top and bottom of the fraction by the same number. Or, if you really don't like reducing, just use your calculator. Divide 4 by 12 to get 0.333. Then divide each answer choice (top by the bottom) to get decimals and see which one matches 0.333. Nice.

2. **F** There are 30 marbles total. $\frac{2}{3}$ are blue. So $\frac{2}{3}$ times the total number of marbles equals the amount of blue marbles. $(\frac{2}{3})(30) = 20$, so there are 20 blue marbles. $\frac{1}{6}$ are red, and $(\frac{1}{6})(30) = 5$, so there are 5 red marbles. The rest are yellow. So 30 − 20 − 5 = 5, and there are 5 yellow marbles.

3. **D** Probability = $\frac{want}{total}$. We want to land in the shaded region. The area of the shaded region where we want to land is: total area − unshaded = shaded, or 12 − 7.5 = 4.5. Great review of Donuts (Skill 20, Area shaded equals

area big guy minus area donut hole.) The total area of the figure is given as 12. So the probability of landing in the shaded region is $\frac{want}{total} = \frac{4.5}{12}$. Oddly, that does not match any answers. Do we freak out; do we assume we're wrong? No! See if any of the answers are equivalent to our answer, by either looking at the fractions or just turning them into decimals on your calculator. Choice D is correct since $\frac{4.5}{12} = \frac{9}{24}$ and since $\frac{4.5}{12} = 0.375 = \frac{9}{24}$.

Skill 37 (page 91)

Anything Times Zero Is Zero

1. **D** You can just "Use the Answers" and try each answer choice to see which one works in the equation. Or you can realize that for the two parts, $(x - 2)$ and $(x + 5)$, to multiply to get zero, one of them must equal zero, since the only way to multiply **to get** zero is to multiply **by** zero. So $x - 2 = 0$ or $x + 5 = 0$. Solve for both possibilities:

$x - 2 = 0$	add 2 to both sides
$x = 2$	or
$x + 5 = 0$	subtract 5 from both sides
$x = -5$	

2. **K** You can just "Use the Answers" and try each answer choice to see which one works in the equation. Or you can realize that the two parts, $(2x + 6)$ and $(2x - 3)$, to multiply to get zero, one of them must equal zero, since the only way to multiply **to get** zero is to multiply **by** zero. So $2x + 6 = 0$ or $2x - 3 = 0$. Solve for both possibilities:

$2x + 6 = 0$	subtract 6 from both sides
$2x = -6$	divide by 2
$x = -3$	or
$2x - 3 = 0$	add 3 to both sides
$2x = 3$	divide by 2
$x = \frac{3}{2}$	

3. **B** You can just "Use the Answers" and try each answer choice to see which one works in

the equation. Or you can factor $x^2 + 4x + 3 = (x + 3)(x + 1)$ and realize that for the two parts, $(x + 3)$ and $(x + 1)$, to multiply to get zero, one of them must equal zero, since the only way to multiply **to get** zero is to multiply **by** zero. So $x + 3 = 0$ or $x + 1 = 0$. Solve for both possibilities:

$x + 1 = 0$ subtract 1 from both sides
$x = -1$ or
$x + 3 = 0$ subtract 3 from both sides
$x = -3$

4. **G** First factor $x^2 + 3x - 10 = (x + 5)(x - 2)$. Since $x^2 + 3x - 10 = (x + 5)(x - 2) = 0$, either $(x + 5)$ or $(x - 2)$ equals zero. So set each equal to zero and solve for x. So $x = -5$ or 2. And the sum of $-5 + 2 = -3$.

5. **E** Great springboard review (Skill 29); remember that $a^4 - 9$ is the ACT's favorite kind of factoring. $a^4 - 9 = (a^2 - 3)(a^2 + 3)$. And

$a^2 - 3$ can factor to $(a - \sqrt{3})(a + \sqrt{3})$. All the possible polynomial factors are $(a^2 - 3)$,

$(a^2 + 3)$, $(a - \sqrt{3})$, and $(a + \sqrt{3})$.

Skill 38 (page 93)

$Y = ax^2 + bx + c$

1. **D** When the equation of a parabola is given in the form $y = ax^2 + bx + c$, the c is the y-intercept. So for $y = x^2 + 2x - 3$, -3 is the y intercept, which has coordinates $(0, -3)$, since the x value is 0 and the y value is -3.

2. **G** Rewrite $y - 2 = -(x + 3)^2$ so it matches the form $y = (x - h)^2 + k$. Just add 2 to both sides to get $y = -(x + 3)^2 + 2$. In this form, the vertex or minimum/maximum point is (h, k), so $(-3, 2)$ is the maximum point.

3. **E** Use the graph. When $y = 0$, the graph is at two different x values, one positive and one negative.

4. **F** Our Skill makes this "hard" question easy! For the graph shown, the vertex clearly has a positive x and a positive y value, and the graph

opens down. So we need h and k to be positive, and we need a negative sign in front of $(x - 2)^2$. Answer choice F is the only equation that fits these requirements, $y - 3 = -(x - 2)^2$.

5. **B** Our Skill makes this "hard" question easy! From the graph, we can see that k must be positive, since the graph opens up. And we can see that r must be negative, since there is a negative y intercept. The product of k and r is (positive) (negative) = negative.

Skill 39 (page 95)

Circles

1. **A** The question describes a circle. That's what a circle is, it's all the points in a plane that are a certain distance from a center point. Remember that radius is the distance from the center to the circle, and the diameter is twice that distance, i.e., twice the radius.

2. **H** The center is (h, k) based on the equation $(x - h)^2 + (y - k)^2 = r^2$. So match up the given equation to $(x - h)^2 + (y - k)^2 = r^2$, and determine h and k. For $(x - 8)^2 + (y + 2)^2 = 49$, $h = 8$ and $k = -2$. Notice that we have $(y + 2)$ but the formula is for $(y - k)$, so we switch the sign of 2 to get -2. So, the center is $(8, -2)$.

3. **D** We just need to apply the info we are given into the equation for a circle, which is $(x - h)^2 + (y - k)^2 = r^2$. Since the center is $(5, -3)$ and the radius is 6, $h = 5$, $k = -3$, and $r = 6$. So D is the answer. Notice that we can easily get this question with process of elimination. For example, since $r = 6$, the number to the right of the equals sign, r^2, must equal 36, not 6.

4. **J** When a picture is described but not shown, draw it. The circle has center $(1, 4)$. Since it is tangent to the y axis, our sketch shows that it has a radius of 1. So $h = 1$, $k = 4$, and $r = 1$. Choice J works.

5. **D** When a picture is described but not shown, draw it. The sketch shows that the center must

be (3, 3) and that the radius is therefore 3. So $h = 3$, $k = 3$, and $r = 3$. Choice D works. Watch out for choice B, which looks okay, but has a minus sign instead of a plus sign between the parentheses! That's why you've got to cut out the flash cards at the end of this book and memorize the equation for a circle.

Skill 40 (page 97)

Weird Circle Factoid

1. **E** The formula for the area of a circle is πr^2. Since the diameter is 20, the radius is 10 (radius is half the diameter), and the area $= \pi(10)^2 = 100\pi$.

2. **G** To find the length of arc AB, find the circumference and then take the part of it that arc AB occupies. Circumference equals πd, or $2\pi r$, since $d = 2r$. So circumference $= 2\pi r = 12\pi$. Since angle $ACB = 30$, arc AB is $\left(\frac{30°}{360°}\right)$ of the whole circumference. So the length of arc $AB = \left(\frac{30°}{360°}\right)(12\pi) = \pi$.

3. **C** Sector ABC is a fancy term for the pizza-shaped wedge that has corners A, B, and C. Here's another great example of our no-intimidation strategy. Even if you didn't know what "sector" meant, if you just went with it and made assumptions, you'd probably guess what it is. Based on the diagram, what else could sector ABC mean? So don't be afraid to just go with it. That's one of the things that the ACT is testing—can you roll with it? So roll with it; you can do it!

   To find the area of sector ABC, find the area of the circle and then take the part of it that sector ABC occupies. Area equals πr^2, so area $= \pi(6)^2 = 36\pi$. Since angle $ACB = 30$, sector ABC is $\frac{30°}{360°}$ of the whole area. So the area of sector $ABC = (\frac{30°}{360°})(36\pi) = 3\pi$.

4. **F** Since the ratio of their radii is 2 : 3, we can just use these as potential radii. So their

circumferences are 4π and 6π, and the ratio of their circumferences is $4\pi : 6\pi$, which reduces to 2 : 3.

5. **C** This is a great question to apply our Skills. First, when you see a right triangle on the ACT, try $a^2 + b^2 = c^2$. So $3^2 + 4^2 = c^2$, and $c = 5$. Now we have the diameter of the circle, and the circumference of the circle $= \pi d = 5\pi$. Also, don't get intimidated by "inscribed," which means "drawn within," because the diagram shows us anyway, and you don't even really need to use the term.

Skill 41 (page 99)

Absolute Value

1. **B** "Use the Answers." Try each answer choice in the equation. Choice B is correct, since $|4 - 16| = |-12| = 12$ and $|4 - (-8)| = |12| = 12$.

2. **K** Again, "Use the Answers." Try each answer choice in the question. The one that does not work is our answer. Choice K is correct since $|-16 - 12| = |-28| = 28$ which is not ≤ 20.

3. **B** Yet again, "Use the Answers." Try each answer choice in the question. Choice B is correct, since $|1 + 2(-5)| = |1 - 10| = |-9| = 9$ which is less than 10. All other choices do not work.

4. **G** And yet again, "Use the Answers." Try each answer choice in the question. Choice G is correct, since $|-4| = 4 = -4 + 8$. All other choices do not work.

5. **A** "Make It Real." This question is way too theoretical, so just choose real numbers for the variable. Then it's easy! Let's try a positive and then a negative number. First, let's say $m = 5$. So

   I. $\sqrt{5^2} = \sqrt{25} = 5$

   II. $|-5| = 5$

   III. $-|5| = -5$

When $m = 5$, choices I and II are equal. Next, let's try $m = -4$.

I. $\sqrt{(-4)^2} = \sqrt{16} = 4$

II. $|-(-4)| = 4$

III. $-|(-4)| = -4$

So when $m = -4$, choices I and II are still equal, and choice A is correct.

Skill 42 (page 101)

Sequences

1. **B** Subtract $47 - 14 = 33$ and divide that number by 3, since we need to add some number three times to get from 14 to 47. Now $33/3 = 11$, so 11 is the number we add each time. Don't be fooled by choice A; 11 is not the answer, it is the number we add to get the answer. $14 + 11 = 25$, and $25 + 11 = 36$, and $36 + 11 = 47$. So 25 and 36 are the numbers that fill the two blanks. You could also just "Use the Answers" to get this question; just try the choices and see which one works. "Use the Answers" is one heck of a strategy!

2. **F** Draw this out: _, −5, 12. Since it's an "arithmetic" sequence, we know that we add some number each time to get the next term. So, $-5 + x = 12$. Add 5 to both sides and $x = 17$. So we add 17 to each term to get the next one. Therefore, the missing first term is $x + 17 = -5$. Don't let this be daunting, just subtract 17 from each side to get −22. You could also do this in your head rather than set up the algebra, or you could, of course, "Use the Answers" and just try the choices to see which one works.

3. **B** A geometric sequence is just a superfancy term for a list of numbers where you multiply each member by the same number to arrive at the next member on the list. To find that number, just notice what it is, or divide a term by the one before it. $-2/4 = -0.5$. So −0.5 is the number you multiply each term by.

The next term is $-\frac{1}{2} \times -0.5 = +0.25$, which matches choice B, $0.25 = \frac{1}{4}$.

4. **F** There are several ways to answer this question. The best way is to whip out your calculator and give it an Air Jordan—just do it. Just add up the 20 terms: $12 + 15 + 18 + 21 + 24 + 27 + 30 + \cdots$. This is much easier than people think. It's totally the way to go. I just timed myself doing this method, and adding up the 20 terms took only 36 seconds. You can even take your time to avoid a careless error. And even if your answer is a little off, you'll probably be able to choose the right answer choice. Notice how far apart they are, so even a close estimate will get you the right one. Not too bad for an easy and flawless way to get a "hard" question right! The other two ways are to write out a bunch of the numbers until you see a pattern that you can use to predict the sum, or to memorize the silly arithmetic sequence sum formula: sum = (0.5)(number terms)(2(first term) + (number terms − 1) (difference between terms)). So sum = s = (0.5)(20)(2(12) + 19(3)) = 810.

Skill 43 (page 103)

Fahrenheit/Celsius Conversions

1. **D** When you are given degrees Celsius, just plug in and simplify. Plug in 34 for C:

$F = \frac{9}{5}C + 32$

$F = \frac{9}{5}(34) + 32$

$F = 93.2 < 93$

2. **G** The thermometer read 63 and went up 10 to 73. You can do the algebra or just "Use the Answers," whichever you prefer. To use the answers, plug in 73 for F, and then try each answer for C, to see which choice works. To do the algebra, plug in 73 for F and solve for C:

$$F = \frac{9}{5}C + 32$$

$73 = \frac{9}{5}C + 32 \qquad$ subtract 32 from each side

$41 = \frac{9}{5}C \qquad$ multiply both sides by $\frac{5}{9}$

$22.77 = C \approx 23$

3. **B** You can do the algebra or just "Use the Answers," whichever you prefer. To use the answers, plug 5 in for F, and then try each answer for C, to see which choice works. To do the algebra, plug in 5 for F and solve for C:

$$F = \frac{9}{5}C + 32$$

$5 = \frac{9}{5}C + 32 \qquad$ subtract 32 from each side

$-27 = \frac{9}{5}C \qquad$ multiply both sides by $\frac{5}{9}$

$-15 = C$

Skill 44 (page 105)

Don't Even Think About It! . . . Most Common Careless Errors I

1. **C** Careless Error Buster: Remember to distribute the negative sign!

    $(9x - 3) - (3x + 6) = 9x - 3 - 3x - 6 = 6x - 9 = 3(2x - 3)$

2. **G** Nice function review! $f(-p)$ means plug $-p$ in for x. So, $f(-p) = -2(-p)^3 = -2(-p^3) = 2p^3$. Careless Error Buster: Remember that the 2 is not in the parenthesis and does not get cubed with the p.

3. **A** Plug in $m = -1$ and simplify. $m(2x^2 - 2) = -1(2x^2 - 2) = -2x^2 + 2$. Careless Error Buster: Remember to distribute the negative sign!

4. **J** Plug $x = 3$ into the equation to get $y = \frac{3(a) - 15}{3} = a - 5$, not $a - 15$. Careless Error Buster: Remember to also divide the 15 by 3.

5. **E** Plug $2p$ in for m and then FOIL $(2p + 4)^2$. You can use the algebra trick for FOILing a

binomial if you know it; and if you don't, no sweat, just do it out $(2p + 4)^2 = (2p + 4)$ $(2p + 4) = 4p^2 + 8p + 8p + 16 = 4p^2 + 16p + 16$. Skill 45 Preview: When you FOIL, remember the middle term!

6. **F** Another function review, this is a good day. This question is similar to question #2 above, but more involved. Plug $-2p$ in for x: $f(-2p) = -2p(2(-2p)^2 - 2) = -2p(2(4p^2) - 2) = -2p(8p^2 - 2) = -16p^3 + 4p$. Careless Error Buster: Remember to distribute the negative sign!

Skill 45 (page 107)

Don't Even Think About It! . . . Most Common Careless Errors II

1. **C** $f(4p)$ means plug $4p$ in for x. So $f(4p) = 3(4p)^2 = 3(16p^2) = 48p^2$. Careless Error Buster: Remember that the 4 gets squared before you multiply by 3. Order of operations, my man.

2. **G** Since she ran a quarter (one-fourth) of a mile in 1 minute 45 seconds, to determine her time for a full mile, we multiply by 4. And 1 minute 45 seconds times 4 is not $(1.45)(4) = 5.8$, **it is $(1.75)(4) = 7$**, since 45 minutes is three-quarters (0.75) of an hour.

3. **D** Plug in $3y$ for x in the expression $(x - 2)^2$ to get $(3y - 2)^2 = 9y^2 - 12y + 4$. Careless Error Buster: When you FOIL a binomial, remember the middle term.

4. **K** Plug -4 into the p function: $p(-4) = 2(-4)(-4 - 2) = 48$. Then plug 48 into the r function: $r(48) = = 16$. Careless Error Buster: Follow the order of operations.

5. **B** Most kids have a tough time with this one. The key is that the distance is equal for both trips, so we set the (rate)(time) equal for the way there and the way back:

    (Rate)(time) = (rate)(time)

    Since the total trip was 0.5 hour, if we call one way t, then the return trip we can call $(0.5 - t)$.

Notice that we should use hours, not minutes, since rate is in miles per **hour**.

$$10(t) = (15)(0.5 - t)$$
$$10t = 7.5 - 15t$$
$$25t = 7.5$$
$$t = 0.3 \text{ hour}$$

Now remember, the question asks for how far Josh lives, NOT the time, so $D = RT$ and $D = 10$ mph $(0.3$ hour$) = 3$ miles. Careless Error Buster: Finish the question, don't stop with $t = 0.3$. Remember to ask, "Did I finish the question?"

Skill 46 (page 109)

Misbehaving Numbers: Weird Number Behavior

1. **B** Since $k = 0$, q must equal zero, since $2kp = 2(0)p = 0$. Anything times zero equals zero.

2. **K** Sketch a diagram of a number line and place the answer choices on the line. Only answer choice K is to the left of -1.

3. **A** Easy with "Make It Real!" Since $x < -1$, let's try $x = -3$. $(-3)^2 = 9$, and $(-3)^3 = -27$. At this point you might say, "Aha, I remember that negative numbers squared become positive, but negative numbers cubed remain negative, so x^2 will always be greater than x^3." Or, if that "aha moment" eludes you, just try a few more possibilities until you feel confident of an answer. Any number for x that you try (as long as you follow the rule $x < -1$) will give you the same result, $x^2 > x^3$.

4. **K** The ACT loves to see if you pick up on this weird behavior stuff. Let's just "Make It Real" and see what happens. Let's say $b = (-0.5)$, and try each answer choice looking for the greatest value. **F.** $b^2 = (-0.5)^2 = 0.25$ **G.** $b^3 = (-0.5)^3 = -0.125$ **H.** $\tan b = \tan(-0.5) = $ (using your calculator) -0.008 **J.** $|b| = |-0.5| = 0.5$ **K.** $b^{-1} = (-0.5)^{-1} = -2$ Answer choice K is smallest.

5. **C** "Make It Real!" Let's say $x = -1.5$. Then $x = -1.5$, $x^2 = (-1.5)^2 = +2.25$, and $x^3 = (-1.5)^3 = -3.375$. So the order is $x^3 < x < x^2$. This makes sense, since when you square a negative, it becomes positive, but when you cube a negative, it stays negative.

Skill 47 (page 111)

Logs

1. **E** $\log_n 144 = 2$ means $n^2 = 144$. To solve $n^2 = 144$, take the square root of both sides to get $n = 12$.

2. **F** $\log_2 x = 5$ means $2^5 = x$, so $x = 2 \cdot 2 \cdot 2 \cdot 2 \cdot 2 = 32$.

3. **A** Each expression translates as "the tiny number to the ? equals the big number." So choice A is largest. 5 to the **2** equals 25. $5^2 = 25$. Each other choice has a smaller answer:

   (A) $\log_5 25 = 2$
   (B) $\log_{25} 25 = 1$
   (C) $\log_{25} 5 = 0.5$
   (D) $\log_{625} 25 = 0.5$
   (E) $\log_{125} 5 = 0.33$

4. **K** Since the first two log equations have the same base m, we can add them. When we add logs, we combine by multiplying; and when we subtract logs, we combine by dividing. So these add to $\log_m(np) = a + b$. This looks pretty similar to the $\log_m(np)^3$ in the question, except for the exponent of 3. Since we can move the exponent in front of the log, we can make them match perfectly, to get $3\log_m(np)$. Since we know that $\log_m(np) = a + b$, we can say that $3\log_m(np) = 3(a + b) = 3a + 3b$. Remember springboard: "when you have two equations in a question, look for the way that they relate."

5. **A** When we add logs, we combine by multiplying; and when we subtract logs, we

combine by dividing. This makes sense since those are the same rules for exponents, and logs are really just a form of exponents. So $\log_3 x + \log_3 9x = \log_3(9x^2) = 4$. This translates to $3^4 = 9x^2$. And since 3^4 is 81, we have $81 = 9x^2$, $x^2 = 9$, and $x = 3$.

6. **F** Let's solve each part that we can: $\log_4 16$ means 2, since $4^2 = 16$; and $\log_6 36$ means 2, since $6^2 = 36$. We can simplify the equation to $2 - 2 = \log_8 x$ or $0 = \log_8 x$, which translates to $8^0 = x$, so $x = 1$, since anything to the zero power equals 1. Nice exponent review!

Skill 48 (page 113)

Not So Complex Numbers

1. **C** "Product" means multiply, so $-2i(3i + 2) = -6i^2 - 4i$. Replace i^2 with (-1) to get $-6(-1) - 4i$ which equals $6 - 4i$. Careless Error Buster: Remember to distribute the negative.

2. **H** Great FOIL review. The square of $(i - 2)$ means $(i - 2)(i - 2)$. So just FOIL $(i - 2)(i - 2)$ to get $i^2 - 2i - 2i + 4$ and collect like terms to get $i^2 - 4i + 4$. Normal FOILing and you'd be done, but here there is one last step, the key to complex number questions. Since i^2 actually equals -1, we substitute -1 in for i^2 to get $-1 - 4i + 4$, and we collect like terms to get a final answer of $3 - 4i$. For some reason, it is standard to put the regular number first and the i second.

3. **D** Since $i^2 = -1$, and since i^4 means $(i^2)(i^2)$, we know that $i^4 = (-1)(-1) = 1$.

4. **K** Don't be intimidated. Just multiply tops and bottoms as in normal fraction multiplication. Remember that the bottoms of $(1 - i)$ and $(1 + i)$ get FOILed:

$$\frac{1}{1-i} \cdot \frac{1-i}{1+i} = \frac{1-i}{1+i-i-i^2} = \frac{1-i}{1-i^2} = \frac{1-i}{1-(-1)} = \frac{1-i}{2}$$

5. **D** Ditto question 4, "Don't be intimidated." This is a great review of factoring. You can multiply tops and bottoms just like normal fraction multiplication, or even better you

can remember to cross-cancel first. How do you know to do that? "Springboard," baby! Whenever you see $x^2 - y^2$, factor it, even if it is disguised as $1 - i^2$. Remember that $x^2 - y^2$ is the ACT's favorite kind of factoring, so watch for it. Once you notice it, it's easy to see what to do next. Don't you love our Mantras! So

$$\frac{5i}{1-i} \cdot \frac{1-i^2}{1+i} = \frac{5i}{1-i} \cdot \frac{(1-i)(1+i)}{1+i} = 5i, \text{ since the}$$

$(1 - i)(1 + i)$ on the tops and bottoms cancel out.

Skill 49 (page 116)

How to Think Like a Math Genius I

1. **B** Ratios and proportions (Skill 21). Just set up the proportion. $\frac{3}{12} = \frac{x}{1}$. Remember to set it up so that chicken is on top in both ratios. If does not actually matter if it's on top or bottom, just that it is in the same spot in both. Then cross-multiply and solve for x. $(3)(1) = (12)(x)$. Divide both sides by 12 to get $x = 0.25$.

2. **F** Supereasy Algebra (Skill 2). Plug $m = 3$ and $n = -2$ into the equation and solve. $(2m - n)^2 = (2(3) - (-2))^2 = (6 + 2)^2 = 8^2 = 64$.

3. **E** When you see a proportion, cross-multiply (Skill 21). $35a = 105$, so $a = 3$.

4. **J** Vocab (Skill 10). The factors of 10 are 1, 2, 5, 10. They are the numbers that divide evenly (no decimal) into the number 10.

5. **E** FOIL (Skill 8). $(4x - 3)^2 = (4x - 3)(4x - 3) = 16x^2 - 12x - 12x + 9 = 16x^2 - 24x + 9$.

6. **H** Isosceles triangles (Skill 7) and triangles have 180° (Skill 6). When two sides are equal, the two angles opposite the two sides are also equal. So we need an answer choice that has at least two angles equal. Choices F, H, and K have at least two angles equal, but since we have a triangle, the angles must also add up to 180, and only choice H adds up to 180.

7. **(E)** Mean (Skill 4). Set up the average formula: $\frac{p_1 + p_2}{2} = 20$. Cross-multiply to get $40 = p_1 + p_2$. So the sum of the perimeters is 40.

8. **(J)** 180 degrees in a triangle (Skill 5). Always mark info from the question into the diagram. All the angles shown comprise a triangle and add up to 180. So $78 + 23 + 45 + x = 180$, and $x = 34$.

9. **(B)** Functions (Skill 16). $f(-1)$ means plug in -1 for x in the f function. So $f(-1) = 3x^3 = 3(-1)^3 = -3$. And $g(5) = x - 2 = 5 - 2 = 3$. So $f(-1) + g(5) = -3 + 3 = 0$.

10. **(J)** Ratios and proportions (Skill 21) and perimeter (Skill 19). The sides have the ratio of $2:3:4$. If the shortest side is actually 10, then each side is 5 times the ratio. So the sides are 10, 15, 20, and the perimeter is $10 + 15 + 20 = 45$.

11. **(E)** Make It Real (Skill 18). "Make It Real" turns this "hard" into an "easy." Choose numbers for the variables. Let's say $a = 9$ and $b = 8$. Then

   (A) works since $ab = (9)(8) = 72$ which is a multiple of 12
   (B) flunks since $3a + 4b = (3)(9) + 4(8) = 59$ which is not a multiple of 12
   (C) works since $4a + 3b = (4)(9) + (3)(8) = 60$ which is a multiple of 12

Since this is a "hard," it's a good idea to test a second set of numbers to confirm that choice A and C both still work. They do. So choice E is correct, since (ab) and $(4a + 3b)$ both work.

Skill 50 (page 120)

How to Think Like a Math Genius II

1. **(E)** Exponents (Skill 30) and avoiding careless errors (Skill 44). $2(3x^4)^3 = 2(27x^{12}) = 54x^{12}$. Careless Error Buster: Remember to apply the exponent to the 3 as well as the x^4, and remember to multiply by 2 only after USING the exponent.

2. **(J)** When you have a right triangle, try $a^2 + b^2 = c^2$ (Skill 24). The legs are 7 and 24, so $(7)^2 + (24)^2 = c^2$, and $49 + 576 = c^2$. So $625 = c^2$, and $25 = c$. Remember to plug in the legs (the shorter sides) for a and b and the hypotenuse (the longest side, always opposite the right angle) for c.

3. **(D)** Translation (Skill 27). $34 equals the original price minus 31% of the original price, so $34 = x - 0.31x$.

4. **(K)** Zero Times Anything Equals Zero (Skill 37). You can just "Use the Answers" on this type of question; try the choices and use the process of elimination, eliminating the ones that don't work. That's a quick, reliable, and easy way to get this one. Or you can factor $x^2 - 5x - 14 = 0$ to $(x - 7)(x + 2) = 0$. Since zero times anything is zero, either $x - 7 = 0$ or $x + 2 = 0$. Solve each of these to get $x = 7$ or -2.

5. **(A)** Midpoint formula (Skill 17). Midpoint is really just the average of the x's and the average of the y's. Just plug in what you know and solve for the variable:

$$\left(\frac{x + x}{2}, \frac{y + y}{2} \right) = \left(\frac{2 + x}{2}, \frac{3 + y}{2} \right) = (-3, 5)$$

In other words, $\frac{2 + x}{2} = -$ and $\frac{3 + y}{2} = 5$. When you see a proportion, cross-multiply, and solve to get $x = -8$ and $y = 7$. $(-8, 7)$. You could also get this question, of course, by "Using the Answers," just trying the choices to see which one works.

6. **(F)** Trig (Skill 34). When a picture is described, draw it. That always helps. Use SohCahToa; the smallest side is 6 and the hypotenuse is 10, so $\sin x° = \frac{6}{10} = 0.6$.

7. **(A)** Weird Number Behavior (Skill 46) and "Make It Real" (Skill 18). Without our Skills many kids would get this one wrong. I almost did, just now, as I was writing this solution. It's tempting to say, "Oh, easy, $9m$ must be the biggest." But remember our Weird Number Behavior Skill, the higher the digit of a negative number,

the smaller the number actually is. Let's "Make It Real," and you'll see what I mean. Let's say $m = -3$. Now -3 is greater than $9(-3)$, which equals -27.

8. ⓖ Similar triangles (Skill 26) and special right triangles (Skill 25). As soon as you see a right triangle with a 30 or 60 degree angle, consider special right triangles. A 30-60-90 triangle has sides that fit the pattern x, $x\sqrt{3}$, $2x$. So since triangle PQR has a 60 degree angle and a side of $3\sqrt{3}$, it must have sides of 3, $3\sqrt{3}$, 6. And triangle MNO, since it is similar, will have sides that are proportional to 3, $3\sqrt{3}$, 6. Choice G works since 6, $6\sqrt{3}$, and 12 is proportional (double) to 3, $3\sqrt{3}$, 6.

9. ⓔ Translation (Skill 27). The question translates to: $20 - 4p = $ a negative. You could also say $20 - 4p < 0$. Solve this to get

$20 - 4p < 0$
$-4p < -20$ subtract 20 from both sides
$p > 5$ divide both sides by -4.

Remember to flip the > sign when you divide by a negative! As is so often the case, you could also just "Use the Answers" to find which one works. Choices A, B, and D do not work. Choice C works, but 20 is not the only solution; when you test choice E, with any number more than 5, it also works, so choice E is correct.

10. ⓖ $y = x^2 + bx + c$ (Skill 38). For the equation $y = x^2 + bx + c$, the sign in front of the x^2 term tells us if the U-shaped graph opens up or down, and the c term tells us the y intercept of the graph. The graph shown in the question opens up, so it should not have a negative in front of the x^2; and it has a negative y intercept, so c should be negative. It also must have an x^2 since it is a parabola (a U-shaped graph). That eliminates choices F, J, and K. Choices G and H are left. Choice G is correct since the -1 outside the parenthesis represents the y intercept, whereas the -1 inside the parenthesis relates to a shift of the x values, not y values. If that last

part is confusing, you could also just graph the two choices on your calculator and see which one matches the diagram shown in the question.

11. ⓓ Algebra (Skill 2) and avoid careless errors (Skill 44). Simplify the inequality:

$3(3x - 2) < 6 - (5x - 2)$
$9x - 6 < 6 - 5x + 2$
$14x < 14$
$x < 1$

Careless Error Buster: Remember to distribute the negative sign.
Once again, you could also just "Use the Answers" and see which choice works.

12. ⓖ "Make It Real" (Skill 18) and area/volume (Skill 19). Let's say $l = 4$, $h = 6$, and $w = 8$; then the surface area of the solid is SA $= 2lw + 2lh + 2wh = 2(4)(8) + 2(4)(6) + 2(8)(6) = 64 + 48 + 96 = 208$. When we halve each dimension, we get $l = 2$, $h = 3$, and $w = 4$, so the new surface area is SA $= 2lw + 2lh + 2wh = 2(2)(4) + 2(2)(3) + 2(4)(3) = 16 + 12 + 24 = 52$. So the new surface area is one-quarter the original; the original times 0.25 equals the new surface area. Without "Make It Real" many people would incorrectly pick choice J, since we halved the dimensions.

13. ⓒ Probability (Skill 36), donut area (Skill 20), or use the diagram (Skill 23). Probability $= \frac{\text{want}}{\text{total}}$, so the probability of hitting the shaded region is $\frac{\text{Want}}{\text{Total}} = \frac{\text{shaded area}}{\text{total area}} = \frac{5^2\pi - 3.5^2\pi}{5^2\pi} = 0.51$.

You could also use the diagram and realize that the shaded region being half the area of the larger circle is the only answer choice that makes sense with the diagram.

14. ⓙ Absolute value (Skill 41) and "Use the Answers" (Skill 1). You could solve the inequality statement or just try the choices and see which is the greatest one that works. Choices F, G, H and J work, but choice J is the largest one that works.

Posttest (page 128)

1. **E** Rather than randomly trying numbers, just "Use the Answers"! Divide each choice by 4, 5, and 6. Only choice E divides evenly (no decimal or remander).

2. **G** This "hard" algebra question seems impossible to many students. "Use the Answers" makes it easy! Just try the choices for x and see which one works. Choice G is correct, since $4^{2+1} = 8^2 = 64$.

3. **D** "T in terms of D and R" means "solve for T, use algebra to get T alone." So, just **divide** both sides by R to get T alone, which gives $\frac{D}{R} = T$.

4. **J** Write out the earnings for the five days: 5, 6, 7, 8, 9. Then to find the "mean," which is just a fancy word for average, add them up and divide by 5.

   $5 + 6 + 7 + 8 + 9 = 35 \div 5 = 7$

5. **B** The question tells us that $\angle BFE = 138°$ and the diagram shows us that $\angle BFE$ also equals $6m$, so we know that $138 = 6m$, and therefore $m = 23$. According to the diagram, $\angle BFD = 5m$ and must thusly (I always wanted to say thusly) be $5(23) = 115$.

6. **K** Use the strategies. Mark $x = 25$. Therefore, in the triangle, the missing angle is $180 - 108 - 25 = 47$. And in the pair of parallel lines, there are two kinds of angles, big and little, and a big + a little = 180. Here 47 is a littler angle, and y is a bigger angle, and therefore equals $180 - 47 = 133$.

7. **A** When a picture is described, draw it. When you see a triangle with two sides equal, mark the angles opposite the equal sides as equal. So $\angle R = 34$ also. Then since a triangle has 180 degrees, the third angle of the triangle must be $180 - 34 - 34 = 112$.

8. **J** $(5x - 2)^2 = (5x - 2)(5x - 2) = 25x^2 - 10x - 10x + 4 = 25x^2 - 20x + 4$. Careless Error Buster: When you FOIL, remember the middle term!

9. **C** You can make a list of the prime numbers less than 50, counting down from 50, or you can "Use the Answers." Just make sure to avoid the careless error of choosing 49 as a prime number (it's divisible by 7, so it's not prime). Either way, the three largest prime numbers less than 50 are 41, 43, and 47.

10. **J** The factors of 36 are 1, 2, 3, 4, 6, 9, 12, 18, and 36. You can use our pairs strategy to factor 36, or you can just "Use the Answers" and see which one works. Remember to <u>underline</u> vocab words to avoid a careless error. Choice G lists the prime factors and choice H lists some multiples, but we want <u>factors</u>.

11. **B** Many kids flee this question, but with our Skills it's easy. Reread Skill 11 to see that the lowest common denominator is the least common multiple of 3, 8, and 12. So just "Use the Answers" and try each choice. Choices B, D, and E are multiples of 3, 8, and 12, but choice B is the lowest and is therefore correct.

12. **K** "Use the Answers"! Simply test each point in the equation to see which one works. Point E has coordinates (2, 4), so $2 = 4 - 2$ which works.

13. **A** Parallel lines have equal slopes. The slope of line $y = -2x - 3$ is -2, so the slope of the two points given is $\frac{-3 - p}{5 - (-2)} = -2$ "Use the Answers" or cross-multiply to solve for p: $p = 11$.

14. **H** Solve the equation for y to get $y = -\frac{3}{4}x - \frac{3}{2}$. The slope is $-\frac{3}{4}$, and a line parallel to this line will have the same slope.

15. **B** The graph shows that Reisner won 25 tournaments. Remember that half a symbol is 5 tournaments, not half a tournament. There were a total of 100 tournaments, so Reisner won $\frac{25}{100} = \frac{1}{4} = 0.25$ of them.

16. **K** He has been to 10 classes, so simply plug in 10 for c to get $3(10) + 8 = 38$. Remember that for a function question, $P(10)$ does not mean P times 10—it means "P of (10)" or "the value of P when c is 10."

17. **C** Use the distance formula:
$$\sqrt{(x - x)^2 + (y - y)^2} =$$
$$\sqrt{(-4 - 7)^2 + (3 - (-2))^2} = \sqrt{(-11)^2 + (5)^2} =$$
$$\sqrt{121 + 25} = \sqrt{146}$$

18. **J** Did you freak out and guess? It might look impossible—unless you use "Make It Real"! Then, it's easy! Just choose a number for x, say 3. When $x = 3$, $y = \frac{5}{2 + 3} = \frac{5}{5} = 1$. To see what happens when x gets larger, let's try $x = 10$: $y = \frac{5}{2 + 10} = \frac{5}{12} = 0.416$. And finally, let's try $x = 1000$ to see what happens when it gets really big: $y = \frac{5}{2 + 1000} = \frac{5}{1002} = 0.005$. So, clearly, as x gets very large, y approaches zero. Another case cracked by "Make It Real"!

19. **C** Since the square requires 80 feet of fence, each side must be 20 feet long. Then the area equals (side)(side) = (20)(20) = 400.

20. **H** Fun question. Remember that the area of a shaded region equals area big guy minus area donut hole. The area of the whole rectangle is $6 \times 9 = 54$, and the area of the unshaded triangle is (this is the cool part) Area $= 0.5bh = 0.5(5)(6) = 15$. It looks like the height wasn't given, but it was—it's 6, it's as tall as the rectangle. So the shaded region $= 54 - 15 = 39$.

21. **D** "Numerator" is just a superfancy word for the top of a fraction, and "denominator" is just the bottom. So if the numerator is 4 more than the denominator and the fraction equals $\frac{5}{3}$, we can represent this as $\frac{d + 4}{d} = \frac{5}{3}$ and cross-multiply to solve for d. Now $(d + 4)(3) = (d)(5)$ which equals $3d + 12 = 5d$. Collect like terms

and $2d = 12$ so $d = 6$. Notice that the question does not ask for d, but for the numerator which is $d + 4$; the numerator is 10. Careless Error Buster: Always check to make sure that you've finished the question and didn't stop in a preliminary step.

22. **J** First multiply the 2 in the second matrix to get $2\begin{bmatrix} w\ x \\ y\ z \end{bmatrix} = \begin{bmatrix} 2w\ 2x \\ 2y\ 2z \end{bmatrix}$. Then subtract the matrices by simply subtracting the numbers in each corresponding spot:
$$\begin{bmatrix} p\ q \\ r\ s \end{bmatrix} - \begin{bmatrix} 2w\ 2x \\ 2y\ 2z \end{bmatrix} = \begin{bmatrix} p - 2w\ q - 2x \\ r - 2y\ s - 2z \end{bmatrix}$$

23. **D** When a picture is described, but not shown, draw it! Don't be a hero and try to do this one without a diagram. A diagram helps so much. It shows you what to do next, and it avoids careless errors. As soon as you draw a decent diagram, you immediately see that to be a right triangle, b must equal 5.

24. **F** If the shortest side is x and the lengths of the sides are consecutive integers, then the longer sides can be called $(x + 1)$ and $(x + 2)$. Now, we just plug these into $a^2 + b^2 = c^2$ to get $(x)^2 + (x + 1)^2 = (x + 2)^2$.

25. **B** love this question. It's rated "medium" only because most kids do not know special right triangles. If you take five minutes right now and memorize them, just the info on the page for Skill 25, then this "medium" becomes easy and you gain points! Once you know about special right triangles, choice B is obviously the answer. It follows the pattern $x, x, x\sqrt{2}$. If, for some crazy reason, you have not memorized the pattern, first of all shame on you, you silly slacker, but second of all, you could draw a diagram for each choice and see which look 45-45-90ish.

26. **H** When a picture is described, draw it. This shows you that we have two triangles, which are similar. Set up a proportion: $\frac{2}{3} = \frac{x}{2.5}$. When you see a proportion, cross-multiply: $5 = 3x$. And divide by 3 to get $1.66 \approx 1.7$.

27. **B** Translate English to math. "Is" means = and "what" means x, so $71 = x(1420)$. Divide both sides by 1420 to get $x = 0.05$, or 5%.

28. **F** Translate English to math. "Of" means multiply, so $(4.3)(1.10) + (0.5)(0.82) = 5.14$.

29. **B** Springboard! When you see $x^2 - y^2$, factor it. So $(x - y)(x + y) = 72$, and we are told that $x - y = 8$, so $8(x + y) = 72$ and $x + y = 9$.

30. **H** Great exponents review! "Product" means multiply, and when you multiply, you add the exponents. But the bases must match. The bases, m and b, do not match, so we cannot combine them to get mb^8; instead, $3m^4 \times 2b^4 = 6m^4b^4 = 6(mb)^4$.

31. **B** Since both terms have an x^2, they are matching terms. To add them, we add their coefficients. "Coefficient" is just the fancy vocab term for the number in front of the variable. So $4x^2 + 5x^2 = 9x^2$.

32. **J** Brian spends 3 minutes per posture, so that number goes next to the p. Brian spends 7 minutes on warm-ups, and this number does not change with the number of yoga postures, so that number goes alone. $y = 3x + 7$.

33. **D** When you see an arrangement question, draw a blank for each position, fill in the # of possibilities to fill each position, and multiply. So we have six seats to fill with six students: $6 \times 5 \times 4 \times 3 \times 2 \times 1 = 720$. Six students were available to fill the first seat, and once 1 sat down, there were only 5 available to fill the next seat, and so on.

34. **G** SohCahToa! Sin L means opposite over hypotenuse, so $\sin L = \frac{8}{17}$.

35. **A** Great beyond SohCahToa question! All you have to do is to follow the directions. Even if you barely know trig, you can get this one if you follow the directions in the question! We are told that the law of cosines states that for any triangle with vertices A, B, and C and the sides opposite those vertices with lengths a, b, and c, respectively, $c^2 = a^2 + b^2 - 2ab \cos C$.

So, to get the length of the side marked with the question mark, we call that side c, which makes its opposite angle C. We assign a and b to the other two sides. Then we plug in the appropriate measures to get $c^2 = a^2 + b^2 - 2ab \cos C = 41^2 + 32^2 - 2(41)(32) \cos 67$.

36. **J** Probability $= \frac{\text{want}}{\text{total}}$, so the probability of NOT selecting a white button is $\frac{12}{16}$. This question requires that we reduce the $\frac{12}{16}$ to $\frac{3}{4}$.

No problem. Stop looking at me like that, reducing is easy; just divide the top and bottom of the fraction by the same number. Or, if you really don't like reducing, use your calculator. Divide 12 by 16 to get 0.75. Then divide each answer choice (top by the bottom) to get decimals and see which one matches 0.75. Nice.

37. **A** Factor $x^2 + 2x - 24 = (x - 4)(x + 6) = 0$. Either $(x - 4)$ or $(x + 6)$ must equal zero, so the solutions are 4 and −6, and the product (multiply) is $(4)(-6) = -24$. This question is a nice review of factoring, of "anything times zero is zero," and of the vocab word "product."

38. **F** Rewrite $y - 5 = (x - 4)^2$ so it matches the form $y = (x - h)^2 + k$, by just adding 5 to both sides to get $y = (x - 4)^2 + 5$. In this form, the vertex or minimum/maximum point is (h, k), so $(4, 5)$ is the minimum point.

39. **D** We just need to apply the info we are given to the equation for a circle, which is $(x - h)^2 + (y - k)^2 = r^2$. Since the center is $(4, 9)$ and the radius is 9, $h = 4$, $k = 9$, and $r = 9$. So D is the answer. Notice that we can easily get this question by the process of elimination. For example, since $r = 9$, the number to the right of the equals sign, r^2, must equal 81, not 9. Also, watch out for choice C, which looks good except that we need a plus sign, not a minus sign, between the parentheses. Moral of the story, "Memorize the equation of a circle."

40. **G** Since the ratio of their radii is 2:3, we can just use these as potential radii. So their areas

are 4π and 9π, and the ratio of their areas is $4\pi : 9\pi$, which reduces to $4:9$.

41. **E** Great "Use the Answers" review! Try each answer choice in the equation. Choice E is correct, since $|15 - 5| = |10| = 10$ and $|-5 - 5| = |-10| = 10$.

42. **J** A geometric sequence is just a superfancy term for a list of numbers where you multiply each member by the same number to arrive at the next member on the list. To find that number, just notice what it is or divide a term by the one before it. $-6/3 = -2$. So -2 is the number you multiply each term by. The next term is $12 \times -2 = -24$.

43. **A** You can do the algebra or just "Use the Answers," whichever you prefer. To use the answers, plug in 137 for F, and then try each answer for C, to see which choice works. To do the algebra, plug in 137 for F and solve for C:

$$F = \frac{9}{5}C + 32$$

$$137 = \frac{9}{5}C + 32 \quad \text{subtract 32 from each side}$$

$$105 = \frac{9}{5}C \quad \text{multiply both sides by } \frac{5}{9}$$

$$58.33 = C \approx 58$$

44. **G** Careless Error Buster: Remember to distribute the negative sign! $(5x + 2) - (8x - 5) = 5x + 2 - 8x + 5 = -3x + 7$

45. **D** When $x = -3$, then $y = 2(-3)^2 + 4 = 22$. Careless Error Buster: PEMDAS, remember to square -3 and then multiply by 2.

46. **H** "Anything times zero is zero." To multiply to get zero, either $(x - 2) = 0$ or $(x + 3) = 0$. So $x = 2$ or -3. You could also just "Use the Answers" and try each choice to see which one(s) work(s).

47. **D** $\log_6 216 = x$ means $6^x = 216$. To solve, just "Use the Answers." Using your calculator, it takes only a few seconds; just try each answer in the equation to see which one works. Choice D is correct since $6^3 = 216$.

48. **G** "The product of 3 and $(2i^2 + x)$ equals -9" translates to $3(2i^2 + x) = -9$. Distribute the 3 to get $6i^2 + 3x = -9$. And since $i^2 = -1$, we can plug that into the equation for i^2, to get $6(-1) + 3x = -9$, which simplifies to $3x = -3$, and $x = -1$.

49. **C** The line through the points $(-6, 3)$ and $(5, 8)$ has slope $\frac{3 - 8}{-6 - 5} = \frac{-5}{-11} = \frac{5}{11}$. Since parallel lines have equal slopes, the slope of the line parallel will also be $\frac{5}{11}$.

50. **J** Great review of weird number behavior and "Make It Real." Plug $m = -2$ and a value for n, say $n = 3$, into each answer choice, and look for the greatest result. Choice J is correct because the even exponent cancels out the negative sign, whereas the odd exponent in choice K keeps the negative.

Glossary

30°, 60°, 90° triangle   a triangle with angles that measure 30°, 60°, 90° has sides that measure x, $2x$, and $x\sqrt{3}$.

45°, 45°, 90° triangle   an isosceles right triangle whose sides have lengths x, x, $x\sqrt{2}$.

absolute value $|-3|$ means "the absolute value of –3." Absolute value means, "after you do the math between the bars, ditch the negative sign!"
- $|-3| = 3$
- $|3| = 3$
- $|-3 - 6| = |-9| = 9$

arithmetic mean   a fancy word for "average." It refers to the normal average that you are used to.

arithmetic sequence   a sequence of numbers where a certain number is added to each term to arrive at the next, like 3, 7, 11, 15, 19, ... In this example, the number 4 is added to each term to arrive at the next.

average   the "average" of a list of numbers is found by adding them and dividing by how many there are.

$$\text{Average} = \frac{\text{sum}}{\text{number of items}}$$

bar graphs   graphs that compare the values of several items, such as sales of different toothpastes.

bisect   to cut into two equal parts. An angle bisector cuts an angle into two equal parts, and a segment bisector cuts a segment into two equal parts.

careless errors   these are bad, "mmmkay."

cartesian plane   a fancy term for the normal grid that you graph lines on.

complex numbers   numbers with a regular part and an imaginary part, such as $2 + 2i$.

consecutive even/odd numbers   even or odd numbers in a row such as 4, 6, 8, 10 or 3, 5, 7, 9.

consecutive numbers   numbers in a row such as 7, 8, 9, 10.

constant term   this expression really throws some kids, but it just means a letter in place of number, kinda like a variable, except that it won't vary.

cos x   the ratio, in a right triangle, of the leg adjacent to angle x to the hypotenuse.

cross-multiply   a method of solving proportions on the ACT. To solve for x in $\frac{5}{12} = \frac{x}{40}$, cross-multiply to get $(5)(40) = 12x$ and divide by 12 to get x alone.

different numbers   numbers that are . . . ummm . . . different.

direct variation   "x varies directly with y" means "x times some number gives y." The formula for this relationship is $y = kx$.

distance   to find the distance between two points, use the formula $\sqrt{(x - x)^2 + (y - y)^2}$.

donut   the area of a donut equals the area of the larger circle minus the area of the donut hole.

equilateral triangle   a triangle with all sides equal and all angles 60°.

equiangular triangle   a triangle with all angles equal, which means that they each must equal 60. The triangle is therefore also equilateral.

even/odd   even numbers are 2, 4, 6, 8, . . . ; odd numbers are 1, 3, 5, 7, ...

exponents, laws of
- $n^6 \times n^2 = n^8$
- $\frac{n^6}{n^2} = n^4$
- $(n^6)^2 = n^{12}$
- $n^0 = 1$
- $n^1 = n$
- $n^{-2} = \frac{1}{n^2}$
- $2n^2 + n^2 = 3n^2$
- $2n + n^2$ does not combine.
- $n^{4/3} = \sqrt[3]{n^4}$

factors   numbers that divide into a number evenly (i.e., without a remainder).
Example: The factors of 48 are 1, 2, 3, 4, 6, 8, 12, 16, 24, 48.

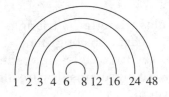

1 2 3 4 6 8 12 16 24 48

FOIL   First, Outside, Inside, Last. Example: $(2x - 5)(5x - 4)$ means multiply $2x$ into the second

parenthesis, multiply −5 into the second parenthesis, and then collect like terms:
$$10x^2 - 8x - 25x + 20 =$$
$$10x^2 - 33x + 20.$$

functions   the cause of much fear in teenagers. A type of equation, like $y = mx + b$. To show that an equation is a function, sometimes people replace the y with $f(x)$ or $g(x)$ or $h(x)$.
$f(x)$ is just a fancy way of saying y. So $f(x) = 2x - 1$ means the same as $y = 2x - 1$.
$f(3)$ means "plug in 3 for x."

geometric sequence   a sequence of numbers where a certain number is multiplied by each term to arrive at the next, like 3, −6, 12, −24, 48, . . . In this example, the number −2 is multiplied by each term to get the next.

greatest common factor   the largest factor shared by several given numbers.
Example: The greatest common factor of 48 and 32 is 16. The number 16 is the largest number that is a factor of both 48 and 16.

integer   a number without decimals or fractions, such as −3, −2, −1, 0, 1, 2, 3, ...

isosceles triangle   a triangle with at least two congruent sides. The two angles opposite the two congruent sides are also congruent.

least common multiple   the lowest number that is a multiple of several numbers.

Example: 60 is the least common multiple of 10, 15, and 20.

logarithm   a fancy way of writing exponents.
$\text{Log}_5 25 = 2$ means
$5^2 = 25$.

lowest common denominator   the lowest number that is a multiple of several denominators.
Example: 60 is the least common denominator for $\frac{3}{10}$, $\frac{7}{15}$, and $\frac{9}{20}$.

line graph   display of how data change, often over time.

linear pair   two adjacent angles that form a line and add up to 180°.

$m + n = 180$

"make it real."   the strategy to use when questions are too theoretical. Choose real numbers in place of the variables, and the question becomes much easier.

math genius   somebody who has mastered all 50 skills from McGraw-Hill's *Top 50 Skills for A Top Score: ACT Math*.

matrix   a chart of data. The ACT might ask you to add, subtract, or multiply matrices.

median   the middle term in a list of numbers. For the list 2, 5, 7, 8, 8, the number 7 is the median.

midpoint   the point halfway between two points.

$\text{Midpoint} = \left(\frac{x + x}{2}, \frac{y + y}{2} \right)$.

mode   the most frequent term in a list of numbers. For the list 2, 5, 7, 8, 8, the number 8 is the mode.

multiples   all the numbers that are divisible by a certain number. The multiples of 3 are 3, 6, 9, 12, 15, 18, 21, etc.

order of difficulty   questions in the ACT math section are arranged from easiest to hardest.

ordered pair   a fancy term for a pair of coordinates (x, y) on the xy coordinate plane.

parallel lines   lines in a plane that never intersect and have equal slopes, like $\frac{2}{3}$ and $\frac{2}{3}$.

perpendicular lines   lines that intersect at a right (90°) angle and have negative reciprocal slopes, like $\frac{2}{3}$ and $-\frac{3}{2}$.

pictographs   graphs that use small pictures to represent data. The key to pictographs is noticing the legend. If each icon represents 8 books, then $1/2$ an icon represents 4 books, not $1/2$ a book.

pie graphs   represent information as part of a pie.

positive/negative numbers   positive numbers are greater than 0 and negative numbers are less than 0

prime factors   the factors of a number that are also prime numbers. The prime factors of 48 are 2 and 3. These are the factors of 48 that also happen to be prime numbers.

prime number a number whose only factors are 1 and itself, such as 2, 3, 5, 7, 11, 13, 17, 19, . . . The number 1 is NOT considered prime, and the number 2 is the only even prime number.

probability a measure of the likelihood of something happening. To determine probability on the ACT, use the equation

$$\text{Probability} = \frac{want}{total}.$$

proportion two ratios set equal to each other, for example, $\frac{5}{12} = \frac{10}{24}$ or $\frac{5}{12} = \frac{x}{40}$. To solve a proportion, cross-multiply.

Pythagorean theorem when given two sides of a right triangle, we can find the third with the formula $a^2 + b^2 = c^2$.

quadratic an expression or equation with a variable, such as x that is squared, for example, $y = ax^2 + bx + c$. In such an equation, the a tells whether the U-shaped graph opens up or down, and the c is the y intercept.

ratio a relationship between two numbers. Such as to bake granola bars, use 7 cups oats to 2 cups sugar. It can be written as 7:2 or $\frac{7}{2}$ or even 7 to 2. The ACT likes to see if you can play with ratios. For example, 4 boys to 5 girls could also be expressed 5 girls to 9 students.

real number any number, such as –3, –2.2, 0, $\sqrt{2}$, π

right triangle a triangle with a right (90°) angle. When you see a right triangle, use $a^2 + b^2 = c^2$ to find the length of a missing side.

similar triangles triangles that have equal angles and proportional sides; one is a shrinky version of the other.

sin x the ratio, in a right triangle, of the leg opposite angle x to the hypotenuse.

slope the measure of a line's steepness. The steeper the line, the bigger the slope. In the equation $y = mx + b$, m is the slope, and b is the y intercept.

$$\text{Slope} = \frac{y_1 - y_2}{x_1 - x_2}$$

special right triangles. a 30°, 60°, 90° triangle with sides x, $2x$, and $x\sqrt{3}$; or a 45°, 45°, 90° triangle with sides x, x, and $x\sqrt{2}$.

springboard strategy on the ACT, when something can be factored, FOILed, reduced, or simplified, do it.

standard (x, y) coordinate plane a fancy term for the normal xy grid that you graph lines on.

tan x the ratio, in a right triangle, of the leg opposite over the leg adjacent to angle x.

translation conversion of word problems from English to math.

transversal a line that crosses two parallel lines, forming 8 angles.

triangle a closed shape formed by three sides. The angles in a triangle add up to 180 degrees.

trigonometry the ways that sides and angles in a triangle are related to each other.

undefined an expression is undefined when it violates math rules, either by having a zero on the bottom of a fraction or by having a negative number in a square root. $\sqrt{-25}$ and $\frac{x-5}{0}$ are undefined.

units digit a fancy term for the "ones" digit in a number, like the 2 in 672.

"use the answers" a terrific strategy to use when you see variables or unknowns in the question and numbers in the answers. Try the answer choices for the variables in the question to see which one works.

vertical angles two angles whose sides form two pairs of opposite rays.

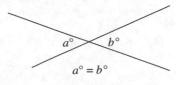

"what is y in terms of x and z" means "solve for y" or "use algebra to get y alone."

x intercept/y intercept the value where a graph crosses the x axis or y axis.˘

Skill 1

When you see **variables** or **unknowns** in the question and **numbers** in the answers . . .

McGraw-Hill's Top 50 Skills for a Top Score: ACT Math by Brian Leaf, M.A.

Skill 3

"What is *m* in terms of *p* and *q*" is just a fancy way of saying . . .

McGraw-Hill's Top 50 Skills for a Top Score: ACT Math by Brian Leaf, M.A.

Skill 4

When you see the word "mean" or "average" on the ACT, use . . .

McGraw-Hill's Top 50 Skills for a Top Score: ACT Math by Brian Leaf, M.A.

Skill 5

When you see vertical angles, a linear pair, or a triangle, . . .

McGraw-Hill's Top 50 Skills for a Top Score: ACT Math by Brian Leaf, M.A.

Here are flash cards for many of our 50 Skills. Cut them out and drill them until you've mastered every one. You can do it!

Skill 6

When you see two parallel lines that are crossed by another line, eight angles are formed and . . .

McGraw-Hill's Top 50 Skills for a Top Score: ACT Math by Brian Leaf, M.A.

When you see **variables** or **unknowns** in the question and **numbers** in the answers, **"Use the Answers."**

"What is m in terms of p and q" is just a fancy way of saying **"solve for m" or "use algebra to get m alone."**

When you see the word "mean" or "average" on the ACT, use $\text{Average} = \dfrac{\text{sum}}{\text{number of items}}$.

When you see vertical angles, a linear pair, or a triangle, **calculate the measures of all angles.**

When you see two parallel lines that are crossed by another line, eight angles are formed and **all of the bigger-looking angles are equal, and all of the smaller-looking angles are equal.**

Skill 7

When you see a triangle with two equal sides, . . .

Skill 8

When you see an expression like $(2x - 5)(5x - 4)$, . . .

Skill 9

Anytime you see a math vocab term, . . .

Skill 10

Define the terms "integer," "prime," "factor," "multiple," "units digit," "consecutive."

Skill 11

How do you find a least common multiple?

When you see a triangle with two equal sides, **mark the two opposite angles as equal, and when all sides of a triangle are equal, mark all angles 60°.**

When you see an expression like $(2x - 5)(5x - 4)$, **multiply $2x$ into the second parenthesis, multiply -5 into the second parenthesis, and collect like terms. $(2x - 5)(5x - 4) = 10x^2 - 8x - 25x + 20 = 10x^2 - 33x + 20$.**

Anytime you see a math vocab term, underline it.

How do you find a least common multiple? **Write out the multiples of the largest number, and then choose the smallest one that is also a multiple of the other numbers. Example: 30 is the LCM of 3, 10, and 15.**

Integer—numbers without decimals or fractions: $-3, -2, -1, 0, 1, 2, 3, \ldots$

Prime—a number whose only factors are 1 and itself. 2, 3, 5, 7, 11, 13, 17, . . . are prime. Note: The number 1 is NOT considered prime, and the number 2 is the only even prime number.

Factors—numbers that divide into a number evenly (i.e., without a remainder)
Example: The factors of 48 are 1, 2, 3, 4, 6, 8, 12, 16, 24, 48.

Multiples—all the numbers that are divisible by a certain number.
Example: The multiples of 3 are 3, 6, 9, 12, 15, 18, 21, etc.

Units digit—just a fancy term for the "ones" digit in a number, like the 2 in 672.

Consecutive numbers—numbers in a row: 7, 8, 9, 10.

Skill 13

To find the slope or "rate of change" of a line, use the formula

Skill 14

Parallel lines have _____ slopes.
Perpendicular lines have _____ slopes.

Skill 17

Midpoint formula =
Distance formula =

Skill 16

f(3) means . . .

Skill 15

The key to understanding charts and graphs is to read the . . .

To find the slope or "rate of change" of a line, use the formula Slope $= \frac{y_1 - y_2}{x_1 - x_2}$

Parallel lines have **equal** slopes, like $\frac{2}{3}$ and $\frac{2}{3}$. Perpendicular lines have **negative reciprocal** slopes, like $\frac{2}{3}$ and $-\frac{3}{2}$.

Midpoint formula $= \left(\frac{x + x}{2}, \frac{y + y}{2} \right)$

Distance formula $= \sqrt{(x - x)^2 + (y - y)^2}$

$f(3)$ means "**plug 3 in for** x."

The key to understanding charts and graphs is to read the **intro material and the "note"** if there is one, and to expect an average, percent, and/or probability question about the data.

Skill 18

When a question with variables is way too theoretical, . . .

McGraw-Hill's Top 50 Skills for a Top Score: ACT Math by Brian Leaf, M.A.

Skill 19

What are the formulas for the area of a triangle, rectangle, square, parallelogram, and circle?

McGraw-Hill's Top 50 Skills for a Top Score: ACT Math by Brian Leaf, M.A.

Skill 20

The area of a donut equals . . .

McGraw-Hill's Top 50 Skills for a Top Score: ACT Math by Brian Leaf, M.A.

Skill 21, card 1

4 boys to 5 girls could also be expressed as . . .

McGraw-Hill's Top 50 Skills for a Top Score: ACT Math by Brian Leaf, M.A.

Skill 21, card 2

When you see a proportion on the ACT, . . .

McGraw-Hill's Top 50 Skills for a Top Score: ACT Math by Brian Leaf, M.A.

When a question with variables is way too theoretical, **"Make It Real."**

---

4 boys to 5 girls could be expressed as:

4 to 5

4 : 5

4/5

4 boys to 9 total students

or 5 girls to 9 total students

Area of triangle = 0.5(base)(height)

Area of rectangle = (base)(height) = (length)(width)

Area of square = (base)(height) = (side)^2

Area of parallelogram = (base)(height)

Area of circle = πr^2

When you see a proportion on the ACT, **cross-multiply.**

The area of a donut equals **the area of the big guy minus the area of the donut hole.**

Skill 22

Tell me about matrices.

Skill 23

When a picture is described but not shown, . . .

Skill 24

When you see a right triangle, try . . .

Skill 25

When you see a 30°, 45°, or 60° angle in a right triangle, try using the . . .

Skill 26

Similar triangles have sides that are . . .

Matrices are really just charts of data, like tables on the sports page. To add matrices, just add numbers that are in corresponding spots. And when multiplying, the result has as many rows as the first matrix and as many columns as the second matrix being multiplied.

When a picture is described, but not shown, draw it! And when a diagram appears not drawn to scale, redraw it.

When you see a right triangle, try $a^2 + b^2 = c^2$.

When you see a 30°, 45°, or 60° angle in a right triangle, try **using the special right triangles.**

Similar triangles have sides that are proportional.

Skill 27

Translate the following from English to math:
Four less than twenty percent of a number is
four.

Skill 29

When something can be factored, FOILed,
reduced, or simplified, . . .

Skill 30

Name the laws of exponents.

Skill 31

$$n^{-3/4} =$$

Skill 32

For the equation $y = mx + b$, the m is
the _____ and b is the _____.

$n^{-3/4} = \frac{1}{\sqrt[4]{n^3}}$

Four less than twenty percent of a number is four. **$0.20x - 4 = 4$**

When something can be factored, FOILed, reduced, or simplified, **do it!**

For the equation $y = mx + b$, the m is the **slope** and b is the **y intercept.**

$n^6 \times n^2 = n^8$ $\frac{n^6}{n^2} = n^4$

$(n^6)^2 = n^{12}$ $n^0 = 1$

$n^1 = n$ $n^{-2} = \frac{1}{n^2}$

$2n^2 + n^2 = 3n^2$ $2n + n^2$ does not combine.

$n^{4/3} = \sqrt[3]{n^4}$

Skill 33

When you see an arrangement question, . . .

Skill 34

What is SohCahToa?

Skill 36

When you see the word "probability," use the equation . . .

Skill 37

What are the solutions to the equation $x^2 - 5x + 4 = 0$?

Skill 38

For the equation $y = ax^2 + bx + c$, the a tells _____, and the c is the _____ .

When you see an arrangement question, **draw a blank for each position, fill in the # of possibilities to fill each position, and multiply.**

What are the solutions to the equation

$x^2 - 5x + 4 = 0$?

$x^2 - 5x + 4 = 0$

$(x - 4)(x - 1) = 0$ factor

so $x = 4$ or 1 set each parenthesis equal to 0

SohCahToa means:

$$\sin = \frac{\text{opposite}}{\text{hypotenuse}} \qquad \cos = \frac{\text{adjacent}}{\text{hypotenuse}}$$

$$\tan = \frac{\text{opposite}}{\text{adjacent}}$$

For the equation $y = ax^2 + bx + c$, the a tells **whether the U-shaped graph opens up or down,** and the c is the y **intercept.**

When you see the word "probability," use the equation Probability $= \frac{\text{want}}{\text{total}}$.

Skill 39, card 1

For the equation of a circle $(x - 3)^2 + (y + 2)^2 = 25$, the center is at point _____ and the radius is _____ units long.

Skill 39, card 2

Name the equation for a circle.

Skill 41

When you see absolute value on the ACT, . . .

Skill 42, card 1

What is the difference between an arithmetic sequence and a geometric sequence?

Skill 42, card 2

What is the next term in the geometric sequence 5, −15, 45, −135, _____?

For the equation of a circle $(x - 3)^2 + (y + 2)^2 = 25$, the center is at point $\underline{(3, -2)}$ and the radius is $\underline{5}$ units long.

The equation for a circle is $(x - h)^2 + (y - k)^2 = r^2$, where (h, k) is the center and r is the radius of the circle.

When you see absolute value on the ACT, **"Use the Answers" or "Make It Real,"** **and remember that absolute value means** **"Ditch the negative sign."**

An arithmetic sequence is a sequence of numbers where a certain number is added to each term to arrive at the next, like 3, 7, 11, 15, 19; and a geometric sequence is a sequence of numbers where a certain number is multiplied by each term to arrive at the next, like 3, −6, 12, −24, 48.

What is the next term in the geometric sequence 5, −15, 45, −135, $\underline{405}$?

Each term is multiplied by −3 to arrive at the next term.

Skill 44

$$\frac{25x^3 - 18}{5x} = ?$$

Skill 45

Careless errors are bad mmmkay, so _____ all vocabulary words and.

Skill 46

Name five weird number behaviors.

Skill 47, card 1

Solve for x.

1. $\log_x 144 = 2$
2. $\log_{10} 10,000 = x$

Skill 47, card 2

$\log_m 30 + \log_m 5 =$

$\log_m 30 - \log_m 5 =$

$\log_m x^2 =$

$$\frac{25x^3 - 18}{5x} = \frac{25x^3 - 18}{5x}$$

This expression does not reduce, since the $5x$ is "under" the $25x^3$ **and** the 18.

1. $\log_x 144 = 2$. So $x = 12$, since $12^2 = 144$.
2. $\log_{10} 10{,}000 = x$. So $x = 4$, since $10^4 = 10{,}000$.

Careless errors are bad mmmkay, so **underline** all vocabulary words and **remember to finish the question**.

$\log_m 30 + \log_m 5 = \log_m (30 \cdot 5) = \log_m 150$

$\log_m 30 - \log_m 5 = \log_m \left(\frac{30}{5}\right) = \log_m 6$

$\log_m x^2 = 2 \log_m x$

1. Small fractions multiplied by small fractions get smaller.
2. The larger the digits of a negative number, the smaller it actually is. Example: $-6 < -1$
3. Subtracting a negative number is like adding. Example: $10 - (-4) = 14$
4. Squaring a negative eliminates it, but cubing does not. Example: $(-3)^2 = 9$, but $(-3)^3 = -27$.
5. Anything times zero equals zero.

Skill 48, card 1

$i =$

McGraw-Hill's Top 50 Skills for a Top Score: ACT Math by Brian Leaf, M.A.

Skill 48, card 2

$(2 - i)(5 + i) =$

McGraw-Hill's Top 50 Skills for a Top Score: ACT Math by Brian Leaf, M.A.

Skill 49

Name at least ten ACT math skills.

McGraw-Hill's Top 50 Skills for a Top Score: ACT Math by Brian Leaf, M.A.

Skill 50, card 1

Name at least ten more ACT math skills.

McGraw-Hill's Top 50 Skills for a Top Score: ACT Math by Brian Leaf, M.A.

Skill 50, card 2

Are you prepared?

McGraw-Hill's Top 50 Skills for a Top Score: ACT Math by Brian Leaf, M.A.

$i = \sqrt{-1}$

$(2 - i)(5 + i)$
$= 10 + 2i - 5i - i^2$
$= 10 - 3i - i^2$
$= 10 - 3i - (-1)$
$= 11 - 3i$

Answers vary; use any of these cards.

Answers vary; use any of these cards.

YES!